INTERNATIONAL CENTRE FOR MECHANICAL SCIENCES

COURSES AND LECTURES - No. 289

MATHEMATICS OF MULTI OBJECTIVE OPTIMIZATION

EDITED BY

P. SERAFINI
UNIVERSITA' DI UDINE

SPRINGER-VERLAG WIEN GMBH

Le spese di stampa di questo volume sono in parte coperte da contributi
del Consiglio Nazionale delle Ricerche.

This volume contains 26 illustrations.

Originally published by Springer Verlag Wien-New York in 1985

ISBN 978-3-211-81860-2 **ISBN 978-3-7091-2822-0 (eBook)**
DOI 10.1007/978-3-7091-2822-0

PREFACE

This volume contains the proceedings of the seminar "Mathematics of Multi Objective Optimization" held at the International Centre for Mechanical Sciences (CISM), Udine, Italy, during the week of September 3-4, 1984.

The seminar aimed at reviewing the fundamental and most advanced mathematical issues in Multi Objective Optimization. This field has been developed mainly in the last twenty years even if its origin can be traced back to Pareto's work. The recent vigorous growth has mainly consisted in a deeper understanding of the process of problem modelling and solving and in the development of many techniques to solve particular problems. However the investigation of the foundations of the subject has not developed at the same pace and a theoretical framework comparable to the one of scalar (i.e. one-objective) optimization is still missing.

It was indeed the purpose of the seminar to review the mathematical apparatus underlying both the theory and the modelling of multi objective problems, in order to discuss and stimulate research on the basic mathematics of the field. The papers of this volume reflect this approach and therefore are not confined only to new original results, but they also try to report the most recent state of the art in each topic.

The contributions in the volume have been grouped in two parts: papers related to the theory and papers related to the modelling of multi objective problems. Then, within each part, the order of the contributions tries, whenever possible, to follow a path from the general to the particular with the minimum discontinuity between adjacent papers. The topics covered in the first part are: value functions both in a deterministic and in a stochastic setting, scalarization, duality, linear programming, dynamic programming and stability; the second part covers comparison of mathematical models, interactive decision making, weight assessment, scalarization models and applications (i.e. compromise and goal programming, etc.).

A particular acknowledgement is due to all the lecturers who contributed so greatly to the success of the seminar with a high intellectual level and clear presentations. Moreover I wish to express my gratitude to all the participants for the pleasant and friendly atmosphere established during the seminar.

The realization of the seminar was made possible by CISM's staff and organizational support, and by the financial contributions of Unesco, Science Sector, and the Committee for Technology and the Committee for Mathematics of the Italian Research Council. To all these Institutions I am deeply grateful.

Paolo Serafini

CONTENTS

PART 2 – MODELLING AND APPLICATIONS

PART 1

THEORY

VALUE FUNCTIONS AND PREFERENCE STRUCTURES

Antonie Stam, Yoon-Ro Lee, and Po-Lung Yu
School of Business
University of Kansas

Abstract

This paper presents existence conditions for a real valued function representation of preference, and the value function representation for special kinds of preference structures. The concepts of additive, monotonic value functions and of preference separability are explored. An efficient method to verify the preference separability condition for the subsets of the criteria index set is discussed, as well as a decomposition algorithm to specify the form of the value function given a collection of preference separable subsets of the index set of criteria. References to techniques for preference elicitation are provided.

Keywords: value function; preference; preference separability; additive and monotonic value functions.

1. Introduction

We shall use the term "value function" rather than "utility function" since in the literature the latter term is commonly associated with decision making under uncertainty (risk).

First some basic concepts and definitions of preference will be studied in section 2. Then (in section 3) we shall state conditions under which a real valued function representation of the preference is feasible. In section 4 we shall explore preference separability and additive and monotonic value functions. In section 5 preference separability will be studied further. Finally, in section 6 we shall present a brief survey of methods which use revealed preference to construct an approximate value function.

2. Concepts of Preference

Let X be the set of alternatives, $f = (f_1, \ldots, f_q)$ the set of criteria, and $Y = \{y = f(x) | x \in X\}$, $Y \subset \mathbb{R}^q$, the outcome space. We shall use superscripts to indicate the elements of Y, and subscripts to indicate the components of y. Thus, y_k^i is the kth component of $y^i \in Y$.

"Preference" can be described by the following: for any two outcomes y^1 and y^2 we shall write $y^1 \succ y^2$ if y^1 is preferred to or better than y^2, $y^1 \prec y^2$ if y^1 is worse than y^2, and $y^1 \sim y^2$ if y^1 is indifferent to y^2 or if the preference relation is indefinite. A preference relation being indefinite implies that it is not known which relation ($\succ$ or $\prec$) actually holds. Note that only one of these three operators can hold for any pair of outcomes. Representing preference by a subset of the Cartesian Product YxY, we have the following:

Definition 2.1 A preference will be one or several of the following: (i) A preference based on $\succ$ (resp. $\prec$ or $\sim$) is a subset of YxY denoted by $\{\succ\}$ (resp. $\{\prec\}$ or $\{\sim\}$), so that whenever $(y^1, y^2) \in \{\succ\}$ (resp. $\{\prec\}$ or $\{\sim\}$), $y^1 \succ y^2$ (resp. $y^1 \prec y^2$, or $y^1 \sim y^2$). (ii) $\{\succsim\} = \{\succ\} \cup \{\sim\}$, and $\{\precsim\} = \{\prec\} \cup \{\sim\}$.

Note that $\{\succ\}$, $\{\prec\}$ and $\{\sim\}$ are known as binary relations.

Example 2.1 Lexicographic Ordering

Let $y = (y_1,\ldots,y_q)$ be such that component y_i is infinitely more important than y_{i+1} $(i=1,\ldots,q-1)$. Then, a lexicographic ordering preference is defined as follows: $y^1 \overset{L}{\succ} y^2$ iff $y_1^1 > y_1^2$, or there exists a k $(2 \leq k \leq q)$ such that $y_k^1 > y_k^2$, and $y_j^1 = y_j^2$ for $j=1,\ldots,k-1$. Observe that $\{\overset{L}{\sim}\} = \{(y,y) | y \in Y\}$.

Example 2.2 Pareto Preference

Assuming without loss of generality that more is better, we define Pareto Preference by $y^1 \overset{P}{\succ} y^2$ iff $y_i^1 \geq y_i^2$ for all $i=1,\ldots,q$ and $y^1 \neq y^2$, i.e., iff $y^1 \geq y^2$. Note that we use the notation $a \geqq b$ $(a,b \in \mathbb{R}^q)$ for "a is greater than or equal to b, componentwise" and $a \geq b$ for "$a_i \geqq b_i$ $i \in \{1,\ldots,q\}$, but not all $a_i = b_i$." The Pareto Preference can thus be described by:

$$\{\overset{P}{\succ}\} = \{(y^1,y^2) | y^1 \geq y^2;\ y^1,y^2 \in Y\};$$
$$\{\overset{P}{\sim}\} = \{(y^1,y^2) | y^1 \text{ not } \geq y^2 \text{ and } y^2 \text{ not } \geq y^1;\ y^1,y^2 \in Y\}.$$

We are now ready to define the better, worse and indefinite/indifferent sets with respect to $y^0 \in Y$:

Definition 2.2

(i) The superior (or better) set with respect to y^0, is defined by $\{y^0 \prec\} = \{y \in Y | y^0 \prec y\}$,

(ii) The inferior (or worse) set with respect to y^0, is defined by $\{y^0 \succ\} = \{y \in Y | y^0 \succ y\}$,

(iii) The indefinite or indifferent set with respect to y^0, is defined by $\{y^0\sim\} = \{y \in Y | y^0 \sim y\}$.

(iv) Given $Y^0 \subset Y$, $\{Y^0 \prec\} = \bigcup_{y^0 \in Y^0} \{y^0 \prec\}$; $\{Y^0 \succ\} = \bigcup_{y^0 \in Y^0} \{y^0 \succ\}$; $\{Y^0 \sim\} = \bigcup_{y^0 \in Y^0} \{y^0 \sim\}$.

If one can describe the preference structure in terms of (i), (ii), (iii) in Definition 2.2 for all $y^0 \in Y$, then one can get the revealed preference. To further investigate the binary relations defined by $\{\succ\}$, $\{\prec\}$ and $\{\sim\}$, we define some properties for the binary relation B on Y:

<u>Definition 2.3</u> The binary relation B on Y is:

(i) <u>symmetric</u> iff $(y^1,y^2) \in B$ implies that $(y^2,y^1) \in B$, for every $y^1,y^2 \in Y$.

(ii) <u>transitive</u> iff $(y^1,y^2) \in B$ and $(y^2,y^3) \in B$ implies $(y^1,y^3) \in B$, for every $y^1,y^2,y^3 \in Y$.

(iii) <u>reflexive</u> if $(y,y) \in B$, for every $y \in Y$.

(iv) <u>complete</u> if $(y^1,y^2) \in B$ or $(y^2,y^1) \in B$, for every $y^1,y^2 \in Y$.

(v) an <u>equivalence</u> if (i), (ii) and (iii) hold.

Note, that the Pareto preference $\{\overset{P}{\succ}\}$ and the lexicographic preference ordering $\{\overset{L}{\succ}\}$ are both transitive. However, $\{\overset{P}{\sim}\}$ and $\{\overset{P}{\succsim}\}$ are not transitive, whereas e.g., $\{y^0\overset{L}{\sim}\}$ consists of only one point, namely y^0 itself. Also, $\{\overset{P}{\succ}\}$ is not complete but $\{\overset{L}{\succ}\}$ is complete.

<u>Definition 2.4</u>

(i) A preference $\{\succ\}$ is a <u>partial order</u> if it is transitive.

(ii) A preference $\{\succ\}$ is a <u>weak order</u> if it is transitive and $\{\succsim\} = \{\succ\} \cup \{\sim\}$ is also transitive.

The following theorems are well known (see e.g., Fishburn,[1] or Yu[2]):

<u>Theorem 2.1</u> If $\{\succ\}$ is transitive and complete then $\{\succsim\}$ is also transitive and complete.

<u>Theorem 2.2</u> Let $\{\succ\}$ be a partial order. Then the preference $\{\succ\}$ is also a weak order <u>iff</u>:

(i) $\{\sim\}$ is an equivalence; and

(ii) if $y^1 \succ y^2$ and $y^2 \sim y^3$, or if $y^1 \sim y^2$ and $y^2 \succ y^3$, then $y^1 \succ y^3$.

We shall see that the preference represented by a value function, discussed in later sections, is a weak order. $\{\overset{P}{\succ}\}$ is a partial order but not a weak order, but $\{\overset{L}{\succ}\}$ is both a partial and a weak order.

Suppose $\{\succ\}$ is a weak order. Then, according to Theorem 2.2(i), $\{\sim\}$ is an equivalence. If we define $\tilde{Y} = \{\tilde{y}\}$ as the <u>collection of all indifference classes $\tilde{y}$</u> in Y, we can define $\tilde{\succ}$ over $\tilde{Y}$ as the preference on the equivalence classes:

<u>Definition 2.5</u> Define <u>$\tilde{\succ}$ over $\tilde{Y}$</u> such that $\tilde{y}^1 \tilde{\succ} \tilde{y}^2$ <u>iff</u> for any $y^1 \in \tilde{y}^1$, $y^2 \in \tilde{y}^2$, we have $y^1 \succ y^2$, where $\tilde{y}^i \in \tilde{Y}$ is the set of points indifferent to y^i.

3. Existence of Value Functions

In this section existence conditions for value functions will be discussed. Many of the results presented have been derived from Fishburn[1]. We shall omit the involved proofs, and emphasize the intuitive concepts. Throughout this and later sections Y will be assumed to be convex. We define a value function as follows:

Definition 3.1 v: $Y \to \mathbb{R}^1$ is called a value function for $\{\succ\}$ on Y if for every $y^1, y^2 \in Y$, we have $y^1 \succ y^2$ iff $v(y^1) > v(y^2)$.

In section 2 we have already introduced the concept of equivalence classes $\tilde{Y}$ of Y. If a value function v -- as defined above -- exists, then we can represent an indifference curve $\tilde{y}$ by an isovalue curve:

$$\tilde{y} = \{y \in Y | v(y) = v^0\}.$$

Important for the existence of a value function v is -- as illustrated by Theorem 3.1 below -- that there exists a countable subset of $\tilde{Y}$ that is $\tilde{\succ}$-dense in $\tilde{Y}$ (see Definition 2.5 for the definition of $\tilde{\succ}$).

Definition 3.2 A subset $A \subseteq Y$ is said to be $\succ$-dense in Y iff $y^1 \succ y^2$, $y^1, y^2 \in Y$, but $y^1, y^2 \notin A$ implies that there exists $z \in A$ such that $y^1 \succ z$ and $z \succ y^2$.

For instance, the rational numbers are "<"-dense in $\mathbb{R}^1$, since for any $r_1 < r_2$, r_1, r_2 not rational, there exists a rational number $r_3 \in \mathbb{R}^1$ such that $r_1 < r_3 < r_2$. The following is a well-known result (see Fishburn,[1] Theorem 3.1):

Theorem 3.1 There exists a value function v for $\{\succ\}$ on Y iff (i) $\{\succ\}$ on Y is a weak order, and (ii) there is a countable subset of $\tilde{Y}$ which is $\tilde{\succ}$-dense in $\tilde{Y}$.

Recall that the Pareto preference on $\mathbb{R}^q$, $q \geq 2$, is not a weak order (because $\{\overset{P}{\sim}\}$ and $\{\overset{P}{\succsim}\}$ are not transitive). Therefore, it cannot be represented by a value function. The lexicographic ordering preference on $\mathbb{R}^q$, $q \geq 2$, cannot be represented by a value function either, even though it is a weak order, because it is complete, so that $\{y^0\sim\} = \{y^0\}$, implying that there exists no countable subset of $\tilde{Y}$ which is $\succ'$-dense in $\tilde{Y}$.

In practice it may be very tedious if not impossible to verify the countability condition of Theorem 3.1. The following theorem compromises some of Theorem 3.1's strength for applicability.

Theorem 3.2 Let Y be a rectangular subset of $\mathbb{R}^q$, such that on Y: (i) $\{\succ\}$ is a weak order; (ii) $y^1 \geq y^2$ implies $y^1 \succ y^2$; (iii) $y^1 \succ y^2$, $y^2 \succ y^3$ implies that there exist $a \in (0,1)$, $b \in (0,1)$ such that $ay^1 + (1-a)y^3 \succ y^2$ and $y^2 \succ by^1 + (1-b)y^3$. Then there exists a value function v for $\{\succ\}$ on Y.

Proof. See Fishburn,[1] Theorem 3.3.

The three conditions (i)-(iii) of Theorem 3.2 are used to construct countable dense subsets of indifference curves in $\tilde{Y}$.

We shall now proceed with some interesting results from topology. For the interested reader who is not familiar with basic definitions of a topology $\mathcal{T}$ for a set Y we have included a short appendix at the end of the paper (Appendix A). Let $\mathcal{T}$ be the topology on Y so that $(Y,\mathcal{T})$ is a topological space. Note that Y is separable iff Y contains a countable subset, the closure of which is Y. The following important result is proved in Debreu[3] and in Fishburn:[1]

Theorem 3.3 There exists a continuous value function v for $\{\succ\}$ on Y in the topology T if: (i) $\{\succ\}$ on Y is a weak order; (ii) (Y,T) is connected and separable; (iii) $\{y\prec\}$, $\{y\succ\} \in T$, for every $y \in Y$.

Comparing Theorem 3.3 with Theorem 3.1, we observe that (i) is identical, while (ii) and (iii) of Theorem 3.3 are used to construct countable dense subsets as needed in (ii) of Theorem 3.1. In addition, (iii) of Theorem 3.3 insures the continuity of v.

In the above we have studied conditions for the existence of a value function. Special properties are usually not assumed. In the following sections (4 and 5) we shall explore the concept of preference separability and the special cases of additive and monotonic value functions.

4. Additive and Monotonic Value Functions and Preference Separability

Assume that there are q different criteria (or attributes), the outcome space Y is the Cartesian product $Y = \prod_{i=1}^{q} Y_i$, where each Y_i is a connected interval in $\mathbb{R}^1$ and let the index set of criteria be $Q = \{1,2,\dots,q\}$. Given a partition of Q: $\{I_1,\dots,I_m\}$, $I_j \neq Q$ $(m \leq q)$, (i.e., $\bigcup_{k=1}^{m} I_k = Q$ and $I_i \cap I_j = \emptyset$ if $i \neq j$), we denote the complement of I_j by $\bar{I}_j = Q \backslash I_j$.

We introduce the following notation:

(i) $z_k = y_{I_k}$ is the vector with $\{y_i | i \in I_k\}$ as its components, $k=1,\dots,m$,

(ii) $Y_{I_k} = \prod_{i \in I_k} Y_i$,

(iii) $y = (y_{I_1},\dots,y_{I_m}) = (z_1,\dots,z_m)$.

Note, that $y_{I_k} \in Y_{I_k}$.

The existence of an additive and monotonic value function for $\{\succ\}$ depends primarily on preference separability:

<u>Definition 4.1</u> Given that $I \subset Q$, $I \neq Q$, $z \in Y_I$ and $w \in Y_{\bar{I}}$, we say that z (or I) is <u>preference separable</u>, or <u>$\succ$-separable</u>, <u>iff</u> $(z^0,w^0) \succ (z^1,w^0)$ for any $z^0,z^1 \in Y_I$ and some $w^0 \in Y_{\bar{I}}$ implies that $(z^0,w) \succ (z^1,w)$ for all $w \in Y_{\bar{I}}$.

Some authors use a different terminology. Keeney and Raiffa[4] for instance use the term preferential independence.

Observe that $\{\overset{P}{\succ}\}$ and $\{\overset{L}{\succ}\}$ are both preference separable with respect to each subset of Q.

Definition 4.1 implies that whenever z is $\succ$-separable, and w is fixed, z^0 is preferred to z^1 no matter where w is fixed, because $(z^0,w^0) \succ (z^1,w^0)$ for any $w^0 \in Y_{\bar{I}}$. This means that if z (or I) is $\succ$-separable we can separate it from the remaining variables in the process of constructing the value function v. However, I being $\succ$-separable does not imply that $\bar{I}$ is $\succ$-separable, even if $\{\succ\}$ does have a value function representation. To illustrate this, consider Example 4.1:

<u>Example 4.1</u> Let $\{\succ\}$ be represented by $v(y) = y_2 \exp(y_1)$, $y_1,y_2 \in \mathbb{R}^1$. We see, that y_2 or $I_2 = \{2\}$ is $\succ$-separable, since $\exp(y_1) > 0$. However, y_2 can be negative, so that y_1 or $\bar{I}_2 = I_1 = \{1\}$ are <u>not</u> $\succ$-separable.

<u>Definition 4.2</u> A value function v(y) is <u>additive</u> <u>iff</u> there are $v_i(y_i)$: $Y_i \to \mathbb{R}^1$, $i=1,\ldots,q$, such that $v(y) = \sum_{i=1}^{q} v_i(y_i)$.

Definition 4.3 If $\{I_1,\ldots,I_m^*\}$ and $z = (z_1,\ldots,z_m)$ are a partition of Q and y respectively, and if $v(z) = (v_1(z_1),\ldots,v_m(z_m))$, then v is said to be strictly increasing in v_i, $i \in \{1,\ldots,m\}$, iff v is strictly increasing in v_i, with v_k $(k=1,\ldots,m;\ k\neq i)$ fixed.

We shall write $v(z) = v(v_k(z_k), \bar{v}_k(\bar{z}_k))$, where $\bar{v}_k(\bar{z}_k)$ denotes the functions of the variables z_j other than z_k (i.e., $j\neq k$), whenever we wish to emphasize z_k. Note that if $v_k(z_k^0) > v_k(z_k^1)$, z_k^0 and $z_k^1 \in Y_{I_k}$, then $(z_k^0, \bar{z}_k) \succ (z_k^1, \bar{z}_k)$ for any fixed $\bar{z}_k$, because of the monotonicity.

The following theorems link additive and monotonic value functions to $\succ$-separability in an obvious way.

Theorem 4.1 If v(y) is additive then $\{\succ\}$ enjoys $\succ$-separability for any subset of Q.

Proof. See Yu,[2] Theorem 5.4.

Theorem 4.2 If v(y) as defined in Definition 4.3 is strictly increasing in v_i, $i \in \{1,\ldots,m\}$, then z_i and I_i are $\succ$-separable.

Proof. See Yu,[2] Theorem 5.5.

The (partial) converse of the above two theorems will be the focus of the remaining discussion of this section.

Definition 4.4 Let $I \subset Q$, $I \neq Q$. I is said to be essential if there exists some $y_{\bar{I}} \in Y_{\bar{I}}$ such that not all elements of Y_I are indifferent at $y_{\bar{I}}$. I is strictly essential if for each $y_{\bar{I}} \in Y_{\bar{I}}$ not all elements of Y_I are indifferent at $y_{\bar{I}}$. If I is not essential it is called inessential.

Obviously, to be essential, Y_I must consist of at least two points. If I is inessential then it does not need to be considered when the value function is constructed. It is therefore innocent to assume that each $i \in Q$ is essential.

We shall now provide several theorems, most of which are due to Gorman[5] and Debreu.[6] These theorems partially reverse Theorems 4.1 and 4.2. We first introduce:

Assumption 4.1

(i) Each topological space (Y_i, T_i) $i=1,\ldots,q$ -- and thus (Y,T), with $Y = \prod_{i=1}^{q} Y_i$, $T = \prod_{i=1}^{q} T_i$ -- is topologically separable and connected;

(ii) $\{\succ\}$ on Y is a weak order, and for each $y \in Y$, $\{y\succ\}$ and $\{y\prec\} \in T$.

If Assumption 4.1 holds, then the existence of a continuous value function v is guaranteed by Theorem 3.3. Theorems 4.3-4.6 will give us a more precise specification of the form of v. First, we shall consider special cases (Theorems 4.3 and 4.4), as discussed by Debreu,[6] which we shall subsequently present in a generalized form (Theorems 4.5 and 4.6), as proposed by Gorman (see Gorman,[5] Lemma 1 and Lemma 2).

Theorem 4.3 Assume that Assumption 4.1 holds. v(y) can be written as $v(y) = F(v_1(y_1),\ldots,v_q(y_q))$, where F is continuous and strictly increasing in v_i $(i=1,\ldots,q)$ which are all continuous, iff each $\{i\}$, $i=1,\ldots,q$ is $\succ$-separable.

Proof. See Debreu.[6]

<u>Theorem 4.4</u> Assume that Assumption 4.1 holds. If there are at least three components of Q that are essential, then we can write $v(y) = \sum_{i=1}^{q} v_i(y_i)$, where each v_i is continuous, <u>iff</u> each possible subset $I \subset Q$ is $\succ$-separable.

Proof. See Debreu.[6]

Together, Theorems 4.1-4.4 illustrate the important role which preference separability plays in determining the additive and monotonic forms of value functions representing $\{\succ\}$.

<u>Theorem 4.5</u> Let $\{I_0, I_1, \ldots, I_m\}$ and $(z_0, z_1, \ldots, z_m)$ be a partition of Q and y respectively. Assume that Assumption 4.1 holds. Then:

$$v(y) = F(z_0, v_1(z_1), \ldots, v_m(z_m)), \tag{4.1}$$

where $F(z_0, \cdot)$ is continuous and strictly increasing in v_i $(i=1,\ldots,m)$, <u>iff</u> each I_i, $i=1,\ldots,m$ is $\succ$-separable.

Proof. See Gorman,[5] p.387, Lemma 1.

Note that if $I_0 = \emptyset$, and each z_i contains only one y_i and q=m, then Theorem 4.5 reduces to Theorem 4.3.

<u>Theorem 4.6</u> Let $I^* = \{I_1, \ldots, I_m\}$ and $(z_1, \ldots, z_m)$ be a partition of Q and y respectively. Assume that Assumption 4.1 holds, $m \geq 3$, and let $\{i\}$ be strictly essential for each $i \in Q$. Then we can write:

$$v(y) = \sum_{i=1}^{m} v_i(z_i) \tag{4.2}$$

<u>iff</u> $\bigcup_{k \in S} I_k$, $S \subset M = \{1,\ldots,m\}$ (i.e., the union of any subsets of $\mathcal{I}^*$) is $\succ$-separable.

Proof. See Gorman,[5] p.388, Lemma 2.

Observe that if $\mathcal{I}^* = \{\{i\} | i \in Q\}$, Theorem 4.6 reduces to Theorem 4.4.

From the above we may conclude that for a value function to be additive, as has been assumed in many applications, we need $\succ$-separability for every subset of Q. In real life a slight violation of this condition may be unavoidable. As long as the violation is not serious, the assumption of additivity of v may not have disastrous consequences, but in each case the sensitivity of the assumptions to violations of the conditions should be studied carefully.

5. Structures of Preference Separability and Forms of Value Functions

This section is devoted to a discussion of a convenient method to verify the $\succ$-separability condition for the subsets of Q. Without an efficient method, one may have to check $\succ$-separability for all 2^q-1 subsets of Q. The method developed below greatly reduces this task. Also, a method is provided to determine the functional form of v(y) given the $\succ$-separability of a collection $\mathcal{I}$ of subsets of Q revealed by the Decision Maker (DM).

<u>Definition 5.1</u> Let $I_1, I_2 \subset Q$. Then I_1 and I_2 are said to <u>overlap</u> <u>iff</u> none of the following: $I_1 \cap I_2$, $I_1 \backslash I_2$, $I_2 \backslash I_1$ are empty.

Example 5.1 Let $Q = \{1,2,3,4\}$, and $\mathcal{I} = \{I_1,I_2,I_3,I_4\}$, where $I_1 = \{1,2\}$, $I_2 = \{2,3\}$, $I_3 = \{1,4\}$, $I_4 = \{4\}$. Then, I_1 and I_2 overlap, as do I_1 and I_3, but I_2 and I_3 do not overlap ($I_2 \cap I_3 = \emptyset$), and I_3, I_4 do not overlap either ($I_4 \setminus I_3 = \emptyset$).

The following Theorem, due to Gorman,[5] p.369, Theorem 1, is a special case, which will later be generalized in Theorem 5.2.

Theorem 5.1 Assume that: (i) Assumption 4.1 is satisfied; (ii) I_1, $I_2 \subsetneq Q$ overlap; (iii) $I_1 \setminus I_2$ or $I_2 \setminus I_1$ is strictly essential; (iv) $\{i\} \subset Q$ is essential, $i=1,\ldots,q$; (v) I_1 and I_2 are $\succ$-separable. Then $I_1 \cup I_2$, $I_1 \cap I_2$, $I_1 \setminus I_2$, $I_2 \setminus I_1$ and $(I_1 \setminus I_2) \cup (I_2 \setminus I_1)$ are all $\succ$-separable and strictly essential.

Definition 5.2 $\mathcal{I}$ is said to be complete if:

(i) $\emptyset$, $Q \in \mathcal{I}$,

(ii) If I_1, $I_2 \in \mathcal{I}$ overlap, then $I_1 \cup I_2$, $I_1 \cap I_2$, $I_1 \setminus I_2$, $I_2 \setminus I_1$, $(I_1 \setminus I_2) \cup (I_2 \setminus I_1)$ are all in $\mathcal{I}$.

Definition 5.3 $C(\mathcal{I})$, the completion of $\mathcal{I}$, is the intersection of all complete collections containing $\mathcal{I}$.

Note that $C(\mathcal{I})$ contains $\emptyset$, Q, and all subsets of Q that can be generated by repeatedly applying Definition 5.2 and Theorem 5.1.

Example 5.2 In Example 5.1, $\mathcal{I}$ is not complete, since e.g., $I_1 \cup I_2 = \{1,2,3\}$ is not in $\mathcal{I}$. The completion of $\mathcal{I}$ is: $C(\mathcal{I}) = \{\{1\}, \{2\}, \{3\}, \{4\}, \{1,2\}, \{1,3\}, \{1,4\}, \{2,3\}, \{2,4\}, \{3,4\}, \{1,2,3\}, \{1,2,4\}, \{1,3,4\}, \{2,3,4\}, \emptyset, Q\}$. Observe, that in this case all subsets of Q are in $C(\mathcal{I})$.

Definition 5.4 Let $\mathcal{I}$ be a collection of subsets of Q. Then:

(i) $\mathcal{I}$ is said to be connected if for any A, B of $\mathcal{I}$ there is a sequence $\{I_1,\ldots,I_s\}$ of $\mathcal{I}$ such that I_{k-1} overlaps with I_k ($k=2,\ldots,s$), and $I_1 = A$, $I_s = B$.

(ii) $\mathcal{I}$ is $\succ$-separable if each element of $\mathcal{I}$ is $\succ$-separable.

Definition 5.5 I, $I \in \mathcal{I}$, is said to be a top element of $\mathcal{I}$ if $I \neq Q$ and I is not contained by any other element of $\mathcal{I}$, except Q.

Example 5.3 The top elements of $C(\mathcal{I})$ in Example 5.2 are $\{1,2,3\}$, $\{1,2,4\}$ $\{1,3,4\}$ and $\{2,3,4\}$.

Theorem 5.1 can be generalized to:

Theorem 5.2 Assume that (i) Assumption 4.1 is satisfied; (ii) $\mathcal{I}$ is connected and $\succ$-separable; (iii) There exists an overlapping pair of elements I_j, I_k of $\mathcal{I}$ so that $I_j \backslash I_k$ or $I_k \backslash I_j$ is strictly essential; (iv) $\{i\} \in Q$ is essential, $i=1,\ldots,q$. Then $C(\mathcal{I})$ is $\succ$-separable.

Proof. See Gorman,[5] p.383, Theorem 3.

Theorem 5.3 Let $\mathcal{I} = \{I_1,\ldots,I_s\}$ be a connected collection of pairs of elements of Q such that $\bigcup_{k=1}^{s} I_k = Q$. Assume that $q \geq 3$, and (i) Assumption 4.1 holds; (ii) $\mathcal{I}$ is $\succ$-separable; (iii) at least one $\{k\}$, $k \in Q$, is strictly essential; (iv) each $\{k\}$, $k \in Q$, is essential. Then each subset of Q is $\succ$-separable and the preference can be represented by an additive value function in the form $v(y) = \sum_{i=1}^{q} v_i(y_i)$.

Proof. See e.g., Yu,[2] Theorem 5.11.

This theorem is important for verifying whether we have $\succ$-separability for all subsets of Q, as illustrated in Example 5.4.

Example 5.4 $\mathcal{I}$ as defined in Example 5.1 is connected and $\succ$-separable (see Definition 5.4(i),(ii)), because each element of $\mathcal{I}$ is $\succ$-separable. Given the assumptions of Theorem 5.3 hold, we can apply Theorem 5.3. In view of Theorems 4.3, 4.4 and 5.1, we can conclude that $C(\mathcal{I}) = P(Q)$ (the collection of all subsets of Q), see e.g., Yu,[2] and we only need to verify q-1 pairs of elements of Q for $\succ$-separability. In this example q=4, so that instead of verifying $2^4-1=15$ subsets of Q, we need to verify only 3 pairs of elements of Q (e.g., {1,2},{2,3},{1,4} or {1,2},{1,3}{1,4}) for $\succ$-separability.

The following decomposition theorem translates the above into an expression in terms of a value function.

Theorem 5.4 Assume that $C(\mathcal{I})$ is $\succ$-separable for some $\succ$-separable collection $\mathcal{I}$ of subsets of Q, and that Assumption 4.1 holds. There are two possible cases:

Case 1: None of the top elements $\{T_1,\ldots,T_m\}$ overlap. Then $\{T_0,T_1,\ldots,T_m\}$, where $T_0 = Q\backslash \bigcup_{i=1}^{m} T_i$, forms a partition of Q and v(y) can be written in the form of (4.1), and Theorem 4.5 applies.

Case 2: Some of $\{T_1,\ldots,T_m\}$ overlap. Then $\{\bar{T}_1,\ldots,\bar{T}_m\}$ forms a partition of Q, where $\bar{T}_i = Q\backslash T_i$. We can write v(y) in the form of (4.2), and Theorem 4.6 applies if each {i}, $i \in Q$, is strictly essential.

Proof. See Gorman,[5] Section 3.

The above theorem can be applied recursively to subsets of $C(\mathcal{I})$ to get the form of the value function. In each iteration, Q and $\mathcal{I}$ need to be appropriately redefined: Given T_i is the top element of the current iteration, the new "Q", denoted by Q^i, is defined by $Q^i := T_i$ (the top element in $C(\mathcal{I})$), and $\mathcal{I}^i := \{I_k | I_k \subset T_i,\ I_k \in \mathcal{I}\}$, which implies $C(\mathcal{I}^i) := \{I_k | I_k \subset T_i,\ I_k \in C(\mathcal{I})\}$. Note that at each iteration if T_i and T_j overlap then $T_i \cup T_j = Q$, since the union of overlapping top elements, if not equal to Q, would have to be contained in $C(\mathcal{I})$. Let us illustrate the process with the following example:

Example 5.5 Let $Q = \{1,2,\ldots,9\}$ be the index set of the criteria (or attibutes), and let $\mathcal{I} = \{\{1,2\},\{2,3,4\},\{5,6\},\{6,7\}\}$ be the collection of subsets of Q which through interaction with the DM have been determined to be $\succ$-separable. Then $C(\mathcal{I}) = \{\emptyset, Q, \{1\}, \{2\}, \{5\}, \{6\}, \{7\}, \{1,2\}, \{3,4\}, \{5,6\}, \{5,7\}, \{6,7\}, \{1,3,4\}, \{2,3,4\}, \{1,2,3,4\}, \{5,6,7\}\}$ is the completion of $\mathcal{I}$. The top elements of $C(\mathcal{I})$ are $T_1 = \{1,2,3,4\}$, $T_2 = \{5,6,7\}$.

Note that by definition $\emptyset$ and Q are included in $C(\mathcal{I})$, and that $\{8,9\}$ is not contained in $C(\mathcal{I})$. This last fact may be due to the DM not revealing any information at all on the preference structure over $\{8,9\}$, or due to the DM not revealing any information about the $\succ$-separability of $\{8,9\}$. Writing $T_0 = Q \setminus \bigcup_{i=1}^{m} T_i$, we have $T_0 = \{8,9\}$ for this example. Observe that in some cases T_0 may be empty. Assume now that the assumptions of Theorem 5.4 are satisfied. The decomposition can be done as follows. $T_0 = \{8,9\}$ can not be decomposed further. $T_1 = \{1,2,3,4\}$ and T_2

= {5,6,7} do not overlap. Theorem 5.4 asserts that we can write the value function v as:

$$v(y) = F(y_8,y_9,v_1(y_1,y_2,y_3,y_4),v_2(y_5,y_6,y_7)). \tag{5.1}$$

Now, let us further decompose v_1 by applying Theorem 5.4 to T_1. Redefining the index set of criteria as $Q^1 = T_1 = \{1,2,3,4\}$, and $\mathcal{I}^1 = \{\{1,2\},\{2,3,4\}\}$, we obtain $C(\mathcal{I}^1) = \{\emptyset, Q^1, \{1\}, \{2\}, \{1,2\}, \{3,4\}, \{1,3,4\}, \{2,3,4\}\}$, and the top elements of $C(\mathcal{I}^1)$ are $T_{11} = \{1,2\}$, $T_{12} = \{1,3,4\}$ and $T_{13} = \{2,3,4\}$. Since these top elements overlap and $\bar{T}_{11} = \{3,4\}$, $\bar{T}_{12} = \{2\}$, $\bar{T}_{13} = \{1\}$, we have

$$v_1(y_1,y_2,y_3,y_4) = v_{11}(y_1) + v_{12}(y_2) + v_{13}(y_3,y_4) \tag{5.2}$$

It is easily checked that no further decomposition of T_1 is possible. It can also be verified that application of Theorem 5.4 to T_2 leads to:

$$v_2(y_5,y_6,y_7) = y_{21}(y_5) + v_{22}(y_6) + v_{23}(y_7) \tag{5.3}$$

In view of (5.1)-(5.3), v(y) can be written as:

$$v(y) = F(y_8,y_9,v_{11}(y_1)+v_{12}(y_2)+v_{13}(y_3,y_4)\ ,\ y_{21}(y_5)+v_{22}(y_6)+v_{23}(y_7))$$

The recursive decomposition process can be illustrated by a tree diagram (Figure 5.1):

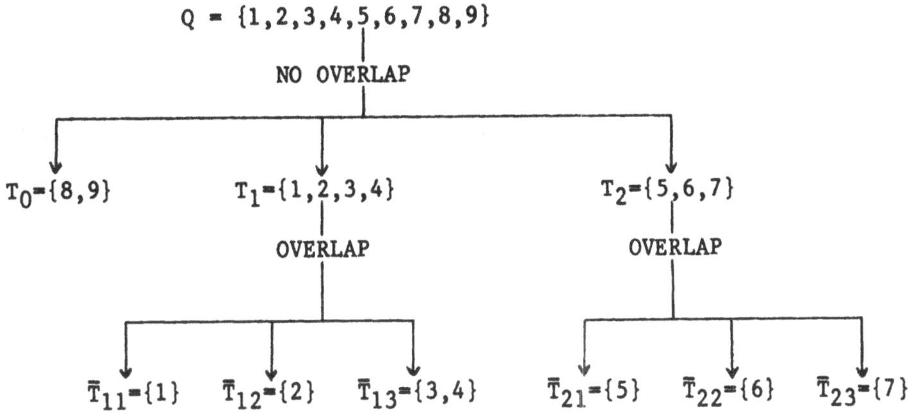

Figure 5.1 Decomposition Tree

6. Survey of Elicitation Techniques for Constructing Value Functions

Although space limitations prevent us from discussing elicitation methods in detail, it is still useful to give a brief summary of frequently used general classes of methods to construct value functions to represent preference information. All methods mentioned below are discussed in detail in Chapter 6 of Yu.[2]

One large class of techniques is a direct application of calculus. These methods are usually based on the construction of approximate indifference curves. This class includes methods using trade-off ratios, tangent planes, gradients and line integrals. Some of these methods are also discussed in Hwang and Masud.[7]

A second class has specifically been developed for the case of additive value functions. A well-known method in this class is the Midvalue Method (Keeney and Raiffa[4]), which is based on pairwise information on $\{\sim\}$ and $\{\succ\}$.

A third class takes into account the fact that usually the revealed preference contains conflicting information, making it a virtually impossible task to construct a consistent value function. The conflicting nature of the information may be due to an unclear perception on the part of the DM of his true preference structure, or imperfect and/or incorrectly used interaction techniques. The objective of these methods is to find a value function and/or ideal point which minimizes the inconsistencies. Some of the techniques, such as regression analysis, are based on statistical theory, others on mathematical programming models, such as least distance and minimal inconsistency methods (using some appropriate ℓ_p-norm). Included are methods based on weight ratios, pairwise preference information, or the distance from a (perhaps unknown) ideal/target point (see Srinivasan and Shocker[8]). For a survey of these methods see, for instance, Hwang and Yoon,[9] and Hwang and Masud.[7] Another group of methods in this class is that of eigen weight vectors (see e.g., Saaty;[10] Cogger and Yu[11]). Yet another group uses holistic assessment. An example is the orthogonal design of experiment (see e.g., Barron and Person;[12] Yu,[2] Section 6.3.3.2).

Each method has its strengths and weaknesses. The selection of the best/correct method is truly an art and poses a challenge for analyst and DM.

Appendix A: Topology

Definition A.1 A topology $\mathcal{T}$ for a set Y is a set of subsets of Y such that:

(i) The empty set ($\emptyset$) and Y are in $\mathcal{T}$.

(ii) The union of arbitrarily many sets of $\mathcal{T}$ is in $\mathcal{T}$.

(iii) The intersection of any finite number of sets of $\mathcal{T}$ is in $\mathcal{T}$.

If $\mathcal{T}$ is a topology for Y, the pair $(Y,\mathcal{T})$ is a topological space. By definition the subsets of Y in $\mathcal{T}$ are called open sets. Let $\mathcal{S}$ be the topology of $\mathbb{R}^1$. Then

Definition A.2 Let $(Y,\mathcal{T})$ be a topological space. Then, $v: Y \to \mathbb{R}^1$ is continuous in $\mathcal{T}$ iff $S \in \mathcal{S}$ implies that $\{y|y \in Y, v(y) \in S\} \in \mathcal{T}$.

Defintion A.3

(i) The closure of $A \subset Y$, $C\ell(A)$, is the set of all $y \in Y$ such that every open set containing y has a nonempty intersection with A.

(ii) A subset B of Y is said to be dense in Y iff $C\ell(B) = Y$.

(iii) $(Y,\mathcal{T})$ is separable iff Y includes a countable subset B such that $C\ell(B) = Y$.

(iv) $(Y,\mathcal{T})$ is connected iff Y cannot be partitioned into two non-empty open sets (in $\mathcal{T}$).

REFERENCES

1. Fishburn, P. C., Utility Theory for Decision Making, John Wiley and Sons, New York, New York, 1970.

2. Yu, P. L., Multiple Criteria Decision Making: Concepts, Techniques and Extensions, Plenum Publishing Company, New York, New York, (Forthcoming).

3. Debreu, G., Representation of a Preference Ordering by a Numerical Function, in Decision Processes, Thrall, R. M., Coombs, C. H. and Davis, R. L., Eds., Wiley, 1954, 159.

4. Keeney, R. L., and Raiffa, H., Decisions with Multiple Objectives: Preferences and Value Tradeoffs, John Wiley and Sons, New York, New York, 1976.

5. Gorman, W. M., The Structure of Utility Functions, Review of Economic Studies, 35, 367, 1968.

6. Debreu, G., Topological Methods in Cardinal Utility, in Mathematical Methods in Social Science, Arrow, K. J., Karlin, S. and Suppes, P., Eds., Stanford University Press, Stanford, California, 1960.

7. Hwang, C. L., and Masud, A. S. M., Multiple Objective Decision Making — Methods and Applications: A State-of-the-Art Survey, Springer-Verlag, New York, New York, 1979.

8. Srinivasan, V., and Shocker, A. D., Estimating the Weights for Multiple Attributes in a Composite Criterion Using Pairwise Judgments, Psychometrika, 38, 473, 1973.

9. Hwang, C. L., and Yoon, K., Multiple Attribute Decision Making -- Methods and Applications: A State-of-the-Art Survey, Springer-Verlag, New York, New York, 1981.

10. Saaty, T. L., A Scaling Method for Priorities in Hierarchical Structures, Journal of Mathemetical Psychology, 15, 234, 1977.

11. Cogger, K. O., and Yu, P. L., Eigen Weight Vectors and Least Distance Approximation for Revealed Preference in Pairwise Weight Ratios, University of Kansas, School of Business, Working Paper, Lawrence, Kansas, 1983.

12. Barron, F. H., and Person, H. B., Assessment of Multiplicative Utility Functions via Holistic Judgments, Organizational Behavior and Human Performance, 24, 147, 1979.

DOMINANCE CONCEPTS IN RANDOM OUTCOMES

Yoo-Ro Lee, Antonie Stam, and Po-Lung Yu
School of Business
University of Kansas

Abstract

As a way to approach decision making under risk or uncertainty, four dominance concepts -- utility dominance, stochastic dominance, mean-variance dominance, and probability dominance -- are reviewed. The characteristic features, the relative merits and shortcomings of these approaches are discussed. Main results and relationships among them are stated. The nondominated set of Ω (a set of random variables) is defined according to different dominance criteria, and interrelationships among them are discussed.

1. Introduction

An essential part of the process of decision making is the selection of the most desirable among possible alternatives. It is very difficult to predict precisely the consequences of each alternative for nontrivial decision problems when the outcomes of each alternative are not known completely. In this situation, each risky alternative may be characterized by a random variable with a probability distribution defined over the possible outcomes, and a decision under risk or uncertainty may be regarded as the selection of random variables.

One way of approaching decision making under risk or uncertainty described above is to incorporate dominance concepts into the decision process. The concept of dominance is concerned with the separation of possible decisions into two sets -- the dominated set and the nondominated set -- by means of a dominance criterion. The main idea is to separate superior alternatives from inferior ones among a given set of alternatives. Superior alternatives can be characterized as nondominated (or efficient) solutions in that they are not dominated by other alternatives.

As random variables can be characterized in many ways, dominance concepts can be studied by many approaches. In this paper we will be concerned with four approaches.

(i) utility dominance -- based on expected utility

(ii) stochastic dominance -- based on cumulative probability distributions

(iii) mean-variance dominance -- based on moments

(iv) probability dominance -- based on outperforming probability

These four approaches are interconnected. In this paper we discuss the characteristic features and the relative merits and shortcomings of these approaches. The main purpose is to investigate results that have already been obtained, and to discuss relationships among them.

In Section 2, notations and some definitions are introduced. In Section 3, utility dominance is discussed with two classes of utility functions, U_1 and U_2. Section 4 is concerned with stochastic dominance. Stochastic dominance in the first degree and second degree are discussed along with examples. The mean-variance dominance concept is discussed in

Section 5. In Section 6, probability dominance is discussed. In Section 7, relationships among the four dominance concepts are investigated and the results are summarized.

2. Notation and Some Definitions

Consider nonempty sets A, S, and C where A is the set of possible alternatives (or actions, decisions), S is the set of states of nature and C is the set of outcomes of the decision problem. We define a random outcome function $f: A \times S \to C$, that determines which outcome will result from a given alternative and states of nature. For convenience, the random outcome function will be denoted by X,Y,..., known as a random variable. Thus, each decision is associated with a random variable. The totality of all random outcome functions or random variables under A will be denoted by Ω. We shall assume that each random variable X or Y has known cumulative probability distribution functions F_X or F_Y where F_X and F_Y are non-decreasing, continuous on the right.

We assume that there exists a real valued function $v: C \to \mathbb{R}^1$ such that for any $c^1, c^2 \in C$, c^1 is preferred to c^2, denoted by $c^1 \succ c^2$, iff $v(c^1) > v(c^2)$. This function is referred to as a value function. A value function does not always exist. However, its existence can be guaranteed under some conditions. See Fisburn,[1] Keeney and Raiffa,[2] Yu[3] and the discussion by Stam, Lee and Yu[4] of the existence conditions for a value function.

In our discussion, we shall assume that the "uncertain outcomes" of a decision can be represented by a one-dimensional random variable whose cumulative distribution is well defined. The outcome is assumed to be "more is better."

3. Utility Dominance

The existence and construction of value functions has been the focus of much theoretical and empirical research in the area of decision making. See Debreu,[5] Gorman,[6] Fishburn,[1] Keeney and Raiffa[2] and Yu.[3] See also Schoemaker[7] for an excellent review of the expected utility model.

In dealing with uncertain outcomes, von Neumann and Morgenstern[8] suggested to construct a real valued function u: $\mathbb{R}^1 \to \mathbb{R}^1$ so that one alternative, represented by random variable X, is preferred to the other alternative, represented by random variable Y iff Eu(X) > Eu(Y). Such a real valued function is known as a utility function for the preference over uncertain outcomes.

Remark 3.1 As noted earlier, much research has been devoted to studying the existence conditions of such a utility function. Undoubtedly the conditions must be extremely strict. To see this point, observe that in general a utility function is an additive weight function defined for the infinite dimensional space in Eu(X) (indexed by t which varies from $-\infty$ to ∞ in $Eu(X) = \int_{-\infty}^{\infty} u(t)\, dF_X(t)$, or $\sum_{t=-\infty}^{\infty} u(t)\, P_X(t)$ when X is discrete and P_X is its mass function). When there is only a finite number of outcomes the weight function can be defined in a finite dimensional space. Note that u(t) is the weight while $dF_X(t)$ or $P_X(t)$ is the measurement of outcomes. For additive weights to exist, strong conditions such as "weak order," "$\succsim$-dense," and very strong "$\succ$-separability" for every subset of the infinite dimension must be imposed. See Stam, Lee, and Yu[4] for a further discussion.

Utility dominance concepts have been studied by many authors. Among them are Hanoch and Levy,[9] Hadar and Russell,[10] and Bawa.[11]

Definition 3.1 Given two random variables X and Y, let u: $\mathbb{R}^1 \to \mathbb{R}^1$ be a utility function. We say that X dominates Y through u, denoted by X u Y iff $Eu(X) > Eu(Y)$.

Remark 3.2 Maximizing Eu(X) over Ω results in a nondominated solution.

In most situations, the acutal construction of the utility function is too complex and unrealistic because complete information about an individual's preference is difficult to obtain. This leads to dominance in a class of utility functions. For instance, although we do not have complete knowledge about u, when it exists, we may know it must be increasing or it must show risk aversion (i.e., people want to pay more than the expected payoff to convert randomness into certainty). In this case dominance can be characterized by using a class of utility functions.

Definition 3.2

(i) Let U_1 be the class of all nondecreasing utility functions. By $X\ U_1\ Y$, we mean $Eu(X) \geq Eu(Y)$ for all $u \in U_1$ and the inequality holds for some $u^0 \in U_1$.

(ii) Let U_2 be the class of all nondecreasing concave utility functions (i.e., U_2 is the subset of U_1 with concave utility functions).

Then, by $X\ U_2\ Y$, we shall mean $Eu(X) \gtrsim Eu(Y)$ for all $u \in U_2$ and the inequality holds strictly for some $u^0 \in U_2$.

Definition 3.3

(i) The nondominated set of Ω with respect to U_1, denoted by $N_1(\Omega,U_1)$ is the collection of all $X \in \Omega$ such that $Y\ U_1\ X$ holds for no $Y \in \Omega$.

(ii) The nondominated set of Ω with respect to U_2, denoted by $N_1(\Omega,U_2)$ is the collection of all $X \in \Omega$ such that $Y\ U_2\ X$ holds for no $Y \in \Omega$.

Remark 3.3 $X\ U_0\ Y$ can be defined for the general class of utility functions, U_0. (U_0 can be linear or quadratic or subject to other specifications.)

4. Stochastic Dominance

Whereas utility dominance is based on utility functions, stochastic dominance is based on the more intuitive monotonicity property of cumulative probability distribution. The concept of stochastic dominance has been studied by Quirk and Saposnik,[12] Fishburn,[13] Hadar and Russell,[10] Hanoch and Levy,[9] and Brumelle and Vickson.[14] We shall summarize some definitions and basic results for stochastic dominance.

Definition 4.1 X stochastically dominates Y in the first degree, denoted by $X\ s_1\ Y$, iff $F_X(a) \lesssim F_Y(a)$ for all $a \in \mathbb{R}^1$, and the strict inequality holds for some $a^0 \in \mathbb{R}^1$.

Remark 4.1 Recall that we assume that the outcome is "more is better." Note that $F_X(a) \leq F_Y(a)$ means $Pr(X > a) = 1-F_X(a) \geq Pr(Y > a) = 1-F_Y(a)$. That is, X has more probability to achieve a goal value "a" than Y. Thus the stochastic dominance is intuitively appealing.

Definition 4.2 X stochastically dominates Y in the second degree, denoted by $X \, s_2 \, Y$ iff $\int_{-\infty}^{a} F_X(t)dt \leq \int_{-\infty}^{a} F_Y(t)dt$ for all $a \in \mathbb{R}^1$ and the strict inequality holds for some $a^0 \in \mathbb{R}^1$.

Example 4.1 Consider the following two probability density functions f(x) and f(y) for random variables X and Y, respectively.

$$f(x) = \begin{cases} 2x, & 0 \leq x \leq 1 \\ 0 & \text{otherwise} \end{cases} \qquad f(y) = \begin{cases} 1 & 0 \leq y \leq 1 \\ 0 & \text{otherwise} \end{cases}$$

Then, cumulative probability distributions are:

$$F_X(x) = \begin{cases} x^2 & 0 \leq x \leq 1 \\ 1 & x > 1 \end{cases} \qquad F_Y(y) = \begin{cases} y & 0 \leq y \leq 1 \\ 1 & y > 1 \end{cases}$$

In this example, X stochastically dominates Y in the first degree as well as in the second degree. Figure 4.1 and Figure 4.2 show $X \, s_1 \, Y$ and $X \, s_2 \, Y$, respectively.

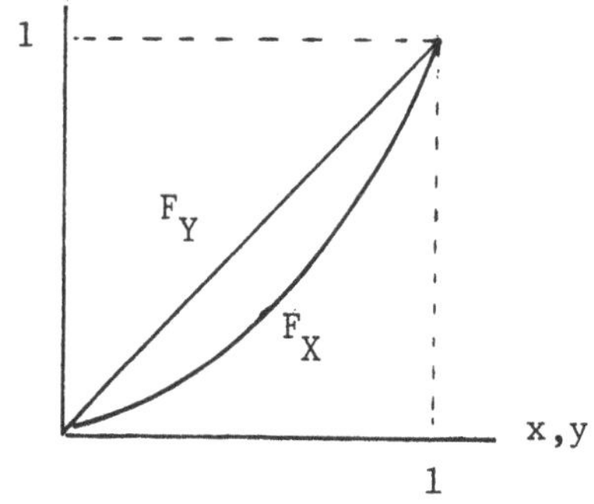

Figure 4.1

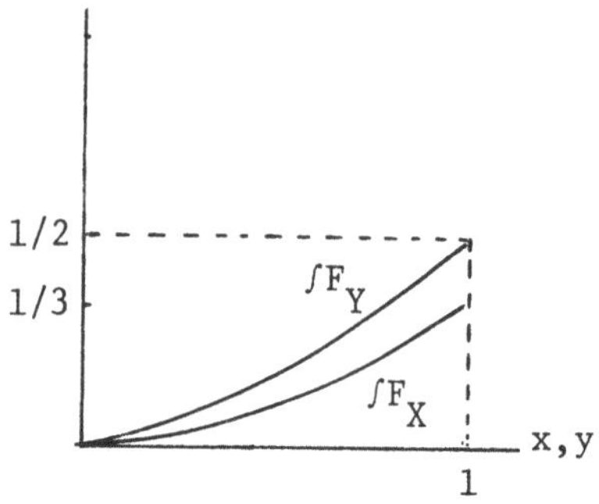

Figure 4.2

Remark 4.2

(i) Graphically (see Figure 4.1), the first degree stochastic dominance means that a probability distribution F_X dominates a probability distribution F_Y iff F_X never lies above F_Y and at least somewhere lies below F_Y.

(ii) Graphically (see Figure 4.2), the second degree stochastic dominance means that F_X dominates F_Y iff the area under F_X never exceeds that under F_Y.

(iii) Clearly X s_1 Y implies X s_2 Y. The converse is not generally true. The following example illustrates this.

Example 4.2 Assume that two random variables X and Y have the following probability distribution.

$$\Pr(X = \tfrac{1}{2}) = \tfrac{2}{3} \qquad \Pr(Y = \tfrac{1}{5}) = \tfrac{1}{2}$$

$$\Pr(X = 1) = \tfrac{1}{3} \qquad \Pr(Y = \tfrac{3}{5}) = \tfrac{1}{2}$$

The cumulative distributions F_X and F_Y are shown in Figure 4.3.

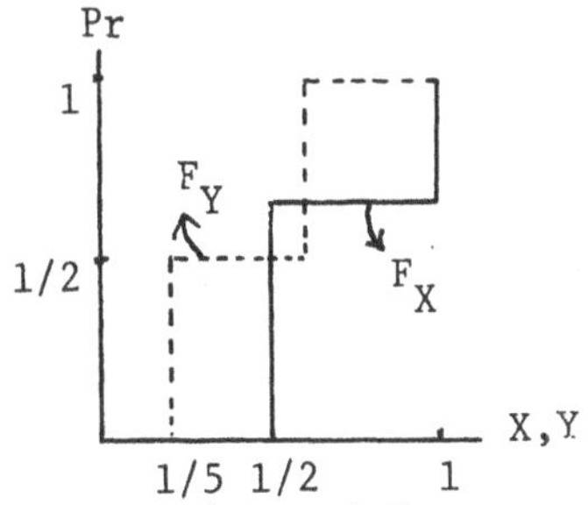

Figure 4.3

As we see in Figure 4.3, X s_2 Y (since the area under F_Y exceeds that under F_X) but not X s_1 Y (since F_X lies above F_Y for some interval).

Remark 4.3 Whenever F_X and F_Y cross over, they cannot stochastically dominate one another in the sense of the first degree. The above example (Example 4.2) shows F_X and F_Y intersect, thus, not X s_1 Y, but X s_2 Y.

Remark 4.4 Note that each random variable is uniquely represented by its cumulative distribution function. Regarding "a" in $F_X(a)$ or $\int_{-\infty}^{a} F_X(t)dt$ as a magnitude of the dimension of "not exceeding a," we see that stochastic dominance is essentially a Pareto preference in the "infinite" dimensional space. Since the dominated cone (see Yu[15] for a discussion) of Pareto preference is $(\frac{1}{2})^{\infty}$ in the entire infinite dimensional space, the set of nondominated random variables will be very large. That is, a large number of random variables will remain nondominated if stochastic dominance, first or second degree, is the only source of preference information. If each distribution is discrete with ℓ possible values (say, \$10, \$20, \$40, \$50 i.e., four possible outcomes), then the dominated cone of Pareto preference will be $(\frac{1}{2})^{\ell}$ of R^{ℓ} (that is, the probability of \$10, the probability of \$20, the probability of \$40, the probability of \$50; each represents a dimension, thus four dimensions).

Definition 4.3 The nondominated set of Ω with respect to s_1, denoted by $N_2(\Omega, s_1)$, is the collection of all $X \in \Omega$ such that $Y s_1 X$ holds for no $Y \in \Omega$.

Definition 4.4 The nondominated set of Ω with respect to s_2, denoted by $N_2(\Omega, s_2)$, is the collection of all $X \in \Omega$ such that $Y s_2 X$ holds for no $Y \in \Omega$.

Remark 4.5 Later we shall show that (i) $N_2(\Omega, s_1)$ includes all the optimal choices of all utility maximizers, and (ii) $N_2(\Omega, s_2)$ includes all the optimal choices of all risk averters.

Remark 4.6 A generalization to s_n, $n \geq 3$ is possible. See Fishburn.[16,17,18]

5. Mean-Variance Dominance

Markowitz[19] and Tobin[20] introduced the mean-variance dominance (or moments dominance) concept for the portfolio selection problems. The mean-variance dominance concept requires the knowledge of the mean and the variance assuming they exist. The basic idea of mean-variance dominance is that a smaller variance (less risky) of the uncertain outcome is preferred to a larger one (riskier) and a larger mean is preferred to a smaller mean. Thus, according to mean-variance dominance, desirable risky alternatives are associated with a smaller variance and larger mean.

Definition 5.1 X dominates Y in the mean-variance sense, denoted by X mv Y, iff $\mu_X \geqq \mu_Y$, $\sigma_X^2 \leqq \sigma_Y^2$ and at least one inequality holds strictly, where $\mu_X = EX$ and $\sigma_X^2 = \text{Var } X$ etc.

Remark 5.1

(i) Mean-variance dominance is a Pareto preference concept after converting the random variable into two measuring criteria (mean and variance).

(ii) Random variables are uniquely characterized by their moments. Some distributions, such as the normal or the gamma distribution, can be uniquely characterized by the first two moments. When we restrict Ω to the class of normal distributions, for instance, mean-variance dominance can provide much information for the final decision.

Definition 5.2 The nondominated set of Ω with respect to mean and variance, denoted by $N_3(\Omega, \text{mv})$ is the collection of all $X \in \Omega$ such that $\mu_Y \geqq \mu_X$ and $\sigma_Y^2 < \sigma_X^2$ or $\mu_Y > \mu_X$ and $\sigma_Y^2 \leqq \sigma_X^2$ hold for no $Y \in \Omega$.

Remark 5.2

(i) Note that the expected utility of quadratic forms is a function of the mean and the variance of the random variables, and when the random variables are normally distributed, their distribution is uniquely determined by the mean and variance. Under these circumstances, problems of the selection of random variables can be determined by the mean and variance and the mean-variance dominance can be powerful. However, quadratic utility functions have been criticized by many authors (see Pratt,[21] Arrow,[22] and Hanoch and Levy[9]). For instance, the assumption of quadratic utility is not appropriate because it indicates increasing absolute risk aversion. Increasing absolute risk aversion means that people become more and more risk averse as the payoff gets larger and larger. The normality assumption may not be appropriate since it rules out asymmetry or skewness in the probability distribution of outcomes.

(ii) Whereas the mean-variance dominance concept has the practical advantage of using information about means and variances only, it is much more restrictive than that of stochastic dominances as it does not take other features of the probability distribution into consideration.

(iii) The mean-variance dominance concept can be naturally extended into dominance in higher moments of the random variables. See Beedles and Simkowitz[23] and Fishburn,[16,17,18] for a further discussion.

6. Probability Dominance

The dominance concepts discussed so far are elaborate methods that are based on a sound theoretical foundation. However, if elaborate methods for the problem of risky choice are too far away from our habitual thinking, they may likely be rejected in practice (see Yu[24] for a discussion). There is a need for developing a method that incorporates the habitual way of making decisions when uncertain outcomes are compared. For this purpose, Wrather and Yu[25] introduced a probability dominance concept, which is based on a habitual way of thinking in making decision.

We first illustrate the rationale for studying probability dominance. Consider the situation in which two uncertain alternatives, represented by random variables X and Y, are evaluated and compared for the decision. Suppose that the following can be said about the two alternatives.

(i) X and Y do not stochastically dominate each other;

(ii) X and Y do not dominate each other in the sense of mean-variance.

From (i) and (ii), we are not able to conclude anything regarding the comparison between X and Y.

(iii) Even though X and Y do not dominate each other in the theoretical sense indicated in (i) and (ii), the decision maker believes that X is likely to outperform Y based on his subjective belief. (See Example 6.1.)

In (iii), the decision maker is able to draw the conclusion that X dominates Y in the probabilistic sense, and (iii) may illustrate a habitual way of thinking of the decision maker.

In the above situation, a habitual way of thinking leads to the result that X dominates Y while elaborate methods indicate that X and Y do not dominate each other. Should we ignore a habitual way of thinking when we make an important decision? It is the contention that we must not overlook habitual methods of reasoning and justifying our decisions. The probability dominance concept is based on this rationale.

Definition 6.1 Given two random outcomes X, Y ε Ω, we say that X dominates Y with probability $\beta \geq 0.5$, denoted by X β Y, iff $Pr(X > Y) \geq \beta$.

Remark 6.1 $Pr(X > Y)$ is the probability that X outperforms Y and β is the likelihood of the outperformance. If $\beta > 0.5$, then $Pr(Y > X) \leq 1 - \beta < 0.5$. Thus, the intuitive meaning of X β Y is that X is likely to be better than Y where "likely" indicates a more than 50-50 chance of occurring.

Example 6.1 Let X and Y be such that $Pr(X = 0.1) = \frac{1}{4}$, $Pr(X = 0.5) = \frac{1}{2}$, $Pr(X = 1) = \frac{1}{4}$ and $Pr(Y = 0.2) = \frac{1}{2}$, $Pr(Y = 0.3) = \frac{1}{2}$. The cumulative distributions F_X and F_Y are shown in the following figure.

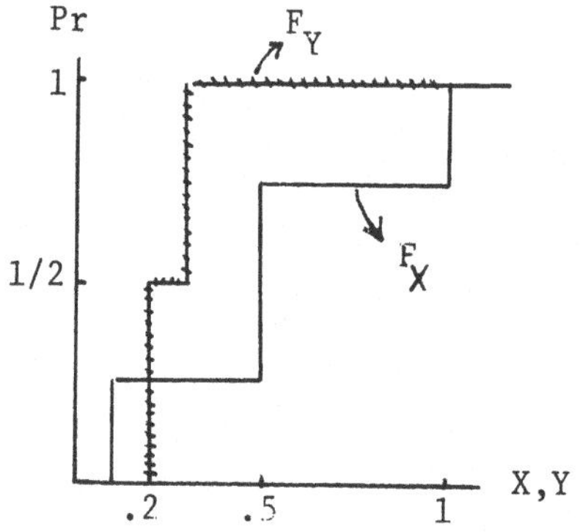

Figure 6.1

In the above example, we see that only 25% of the time X < Y and 75% of the time X > Y. Thus, $Pr(X > Y) = Pr(X = 0.5) + Pr(X = 1) = \frac{3}{4}$. This means $X \beta Y$ for any $\beta \in [0.5, 0.75]$.

Remark 6.2 In the above example we observe the following:

(i) X and Y do not stochastically dominate each other in either the first degree or the second degree.

(ii) X and Y do not dominate each other in the sense of mean-variance.

(iii) However, X dominates Y with probability $\beta \in [0.5, 0.75]$.

Remark 6.3 A weakness of the probability dominance concept is that "inequality" is emphasized in determining the dominance relation and the magnitude of the difference in the inequality is ignored. The following example illustrates the difficulty with the probability dominance concept when used alone.

Example 6.2 Let X and Y be defined as $Pr(X = 0) = 0.3$, $Pr(X = 10) = 0.7$ and $Pr(Y = 9) = 1$. Then $X \beta Y$ with $\beta \in [0.5, 0.7]$. However, intuitively Y is superior to X. In fact $Y s_2 X$ and Y mv X.

In the following we summarize some important properties of probability dominance. These are not generally shared by stochastic or mean-variance dominance (see Wrather and Yu[25] for details).

Definition 6.2 The nondominated set of Ω with respect to β, denoted by $N_4(\Omega, \beta)$, is the collection of all $X \in \Omega$ such that $Y \beta X$ holds for no $Y \in \Omega$.

Theorem 6.1 Let $X, Y \in \Omega$

(i) If $\beta_1 > \beta_2$ and $X \beta_1 Y$ then $X \beta_2 Y$.

(ii) If $\beta_1 > \beta_2$ then $N_4(\Omega, \beta_1) \supset N_4(\Omega, \beta_2)$

(iii) Let $u: \mathbb{R}^1 \to \mathbb{R}^1$ be any strictly increasing function. Then $X \beta Y$ iff $u(X) \beta u(Y)$.

Theorem 6.2 Let Z be a standard normal random variable and z_β be such that $F_Z(z_\beta) = \beta$. Then $X \beta Y$ iff $\mu_Y + z_\beta \sqrt{\sigma_X^2 + \sigma_Y^2} \leqq \mu_X$ holds.

Remark 6.3 One of the shortcomings of probability dominance is that transitivity holds only when some specific conditions are met. When Ω is a collection of mutually independent random variables which have negative exponential distributions or which have normal distributions, then X β Y and Y β Z will imply that X β Z.

7. Some Relationships among Different Dominance Concepts

In this section we investigate relationships among the four dominance concepts discussed in the previous sections. First, we summarize the dominance concepts with respect to different dominance criteria in Figure 7.1.

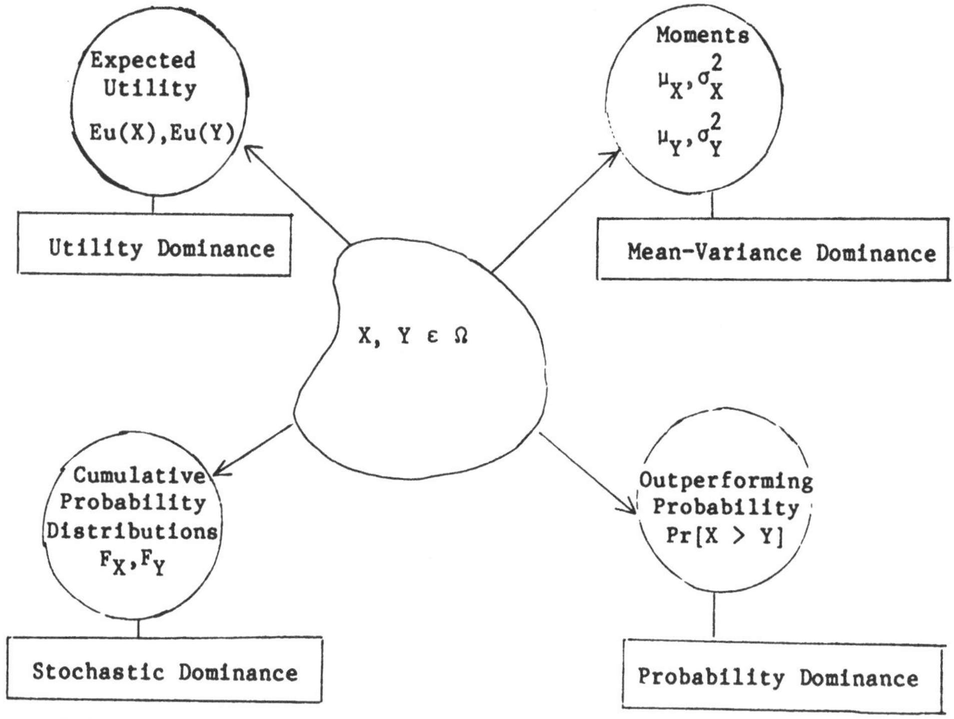

Figure 7.1

Theorem 7.1

(i) X s_1 Y iff X U_1 Y

(ii) X s_2 Y iff X U_2 Y

Proof. See Hadar and Russell,[10,26] Hanoch and Levy,[9] and Bawa.[11]

Remark 7.1 Both $X\ s_i\ Y$ and $X\ U_i\ Y$, $i=1,2$ are essentially Pareto preference in the infinite dimensional space. Theorem 7.1 shows their equivalence. In the infinite dimensional space, the dominated cone of Pareto preference is very small and the set of nondominated alternatives is very large unless Ω is small and/or with some special structures.

Theorem 7.2

(i) Let $F_{Y|X}(t) = \Pr[Y \leq t|X = t]$. Then $E_X[F_{Y|X}(X)] = \int_{-\infty}^{\infty} F_{Y|X}(t)dF_X(t) \geq \beta$ is a necessary condition for $X\ \beta\ Y$. It is also a sufficient condition when $F_{Y|X}$ is continuous on $\mathbb{R}^1$.

(ii) Let X and Y be independent. Then $E_X F_Y(X) \geq \beta$ is a necessary condition for $X\ \beta\ Y$. It is also a sufficient condition when F_Y is continuous on $\mathbb{R}^1$.

Proof. See Wrather and Yu.[25]

Remark 7.2 That X and Y are independent is a resonable assumption, since in the choice model only one of the random variables will be chosen as a desirable choice. When we have to select a combination of random variables, each combination may be regarded as a random variable by redefinition. Note that $F_{Y|X}$ and F_Y are nondecreasing functions (i.e., they are two functionals in U_1). Theorem 7.2 converts probability dominance into a comparison in terms of expected values by regarding $F_{Y|X}$ or F_Y as the functional for the expectation.

Theorem 7.3 Let X and Y be independent, with F_X and F_Y continuous. Then $X\ s_1\ Y$ implies that $X\ \beta\ Y$ for $\beta \geq 0.5$.

Proof. See Wrather and Yu.[25]

Remark 7.3 Theorem 7.3 fails to be true if X and Y are dependent. See Wrather and Yu.[25]

Theorem 7.4 Let X, Y and Z be any normally distributed random variables. Then,

(i) $X \; s_2 \; Y$ iff X mv Y

(ii) $X \; s_2 \; Y$ iff $X \; U_2 \; Y$

(iii) X mv Y and $Y \; \beta \; Z$ imply that $X \; \beta \; Z$

Proof. For (i), see Hanoch and Levy.[9] Proof of (ii) follows from Theorem 7.1. For (iii), see Wrather and Yu.[25]

We have defined nondominated sets according to different dominance criteria. The basic idea of the nondominated set is to narrow the set of all available alternatives down to a set which contains the optimal choice by eliminating the inferior alternatives that are dominated by at least one alternative in the set. The relationships among nondominated sets with respect to the different criteria can be summarized in the following way:

(i) $N_1(\Omega, U_2) = N_2(\Omega, s_2) \subseteq N_2(\Omega, s_1) = N_1(\Omega, U_1)$

(ii) $N_2(\Omega, s_1) \supset N_4(\Omega, \beta)$ if Ω is a class of independent random variables and $\beta \geq 0.5$.

(iii) $N_2(\Omega, s_2) = N_3(\Omega, mv)$ if Ω is a class of normally distributed random variables.

In the following figure we summarize relationships among the four dominance concepts.

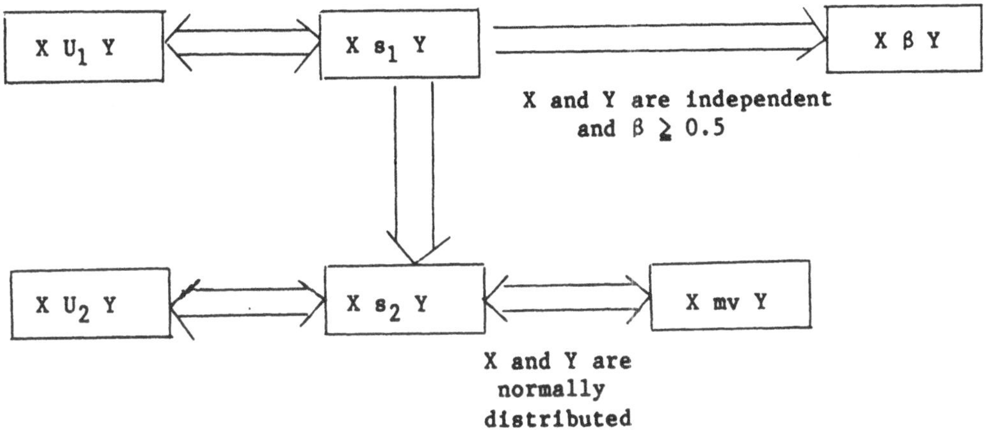

Figure 7.2

References

1. Fishburn, P. C., Utility Theory for Decision Making, John Wiley and Sons, New York, New York, 1970.

2. Keeney, R. L., and Raiffa, H., Decisions with Multiple Objectives: Preferences and Value Tradeoff, John Wiley and Sons, New York, New York, 1976.

3. Yu, P. L., Multiple Criteria Decision Making: Concepts, Techniques and Extensions, Plenum Press, New York, New York, (Forthcoming).

4. Stam, A., Lee, Y. R., and Yu, P. L., Value Functions and Preference Structures, in Proceedings of International Seminar on Mathematics of Multi Objective Optimization, CISM, Udine, Italy, 1984.

5. Debreu, G., Topological Methods in Cardinal Utility, in Mathematical Methods in Social Science, Arrow, K. J., Karlin, S. and Suppes, P., Eds., Stanford University Press, Stanford, California, 1960.

6. Gorman, W. M., The Structure of Utility Functions, Review of Economic Studies, 35, 367, 1968.

7. Schoemaker, P. J., The Expected Utility Model: Its Variants, Purposes, Evidence and Limitations, Economic Literature, 20, 529, 1982.

8. Von Neumann, J., and Morgenstern, O., Theory of Games and Economic Behavior, 2nd edition, Princeton University Press, Princeton, New Jersey, 1947.

9. Hanoch, G., and Levy, H., The Efficiency Analysis of Choices Involving Risk, Review of Economic Studies, 36, 335, 1969.

10. Hadar, J., and Russell, W. R., Rules for Ordering Uncertain Prospects, American Economic Review, 59, 25, 1969.

11. Bawa, V. S., Optimal Rules for Ordering Uncertain Prospects, Journal of Financial Economics, 2, 95, 1975.

12. Quirk, J. P., and Saposnik, R., Admissibility and Measurable Utility Functions, Review of Economic Studies, 29, 140, 1962.

13. Fishburn, P. C., Stochastic Dominance: Theory and Application, in The Role and Effectiveness of Theories of Decision in Practice, White, D. J. and Bowen, K. C., Eds., Hodder and Stoughton, London, 1975.

14. Brumelle, S. L., and Vickson, R. G., A Unified Approach to Stochastic Dominance, in Stochastic Optimization Models in Finance, Ziemba, W. T. and Vickson, R. G., Eds., Academic Press, New York, New York, 1975.

15. Yu, P. L., Cone Convexity, Cone Extreme Points and Nondominated Solutions in Decision Problems with Multiobjectives, Journal of Optimization Theory and Applications, 14, 319, 1974.

16. Fishburn, P. C., Stochastic Dominance and Moments of Distributions, Mathematics of Operations Research, 5, 94, 1980.

17. Fishburn, P. C., Continua of Stochastic Dominance Relations for Unbounded Probability Distributions, Journal of Mathematical Economics, 7, 271, 1980.

18. Fishburn, P. C., Moment-Preserving Shifts and Stochastic Dominance, Mathematics of Operations Research, 7, 629, 1982.

19. Markowitz, H. M., Portfolio Selection, Journal of Finance, 6, 77, 1952.

20. Tobin, J. E., Liquidity Preference as Behavior Towards Risk, Review of Economic Studies, 25, 65, 1958.

21. Pratt, J. W., Risk Aversion in the Small and in the Large, Econometrica, 32, 122, 1964.

22. Arrow, K. J., Essays in the Theory of Risk-Bearing, Markham, Chicago, Illinois, 1971.

23. Beedles, W. L., and Simkowitz, M. K., A Note on Skewness and Data Errors, Journal of Finance, 33, 288, 1978.

24. Yu, P. L., Behavior Bases and Habitual Domains of Human Decision/Behavior: An Integration of Psychology, Optimization Theory and Common Wisdom, International Journal of Systems, Measurement, and Decisions, 1, 39, 1981.

25. Wrather, C., and Yu, P. L., Probability Dominance in Random Outcomes, Journal of Optimization Theory and Applications, 36, 315, 1982.

26. Hadar, J. and Russell, W. R., Stochastic Dominance and Diversification, Journal of Economic Theory, 3, 288, 1971.

SCALARIZATION IN MULTI OBJECTIVE OPTIMIZATION

Johannes Jahn
Department of Mathematics
Technical University of Darmstadt

ABSTRACT

In this paper general multi objective optimization problems are investigated for different optimality notions. For these problems appropriate single objective optimization problems are presented whose optimal solutions are also optimal for the multi objective optimization problem. And conversely, for optimal solutions of a multi objective optimization problem suitable single objective optimization problems are considered which have the same optima. These results lead even to a complete characterization of the optimal solutions of multi objective optimization problems. Finally, this theory is applied to vector approximation problems.

1. INTRODUCTION

In general, scalarization means the replacement of a multi objective optimization problem by a suitable single objective optimization problem which is an optimization problem with a real-valued objective functional. It is a fundamental principle in multi objective optimization that optimal solutions of these problems can be characterized as solutions of certain single objective optimization problems. Since the single objective optimization theory is widely developed scalarization turns out to be of great importance for the multi objective optimization theory.

It is the aim of this paper to present some necessary and sufficient conditions for optimal solutions of general multi objective optimization problems via scalarization. This theory (which differs from the approach of Pascoletti/Serafini[31]) is developed for different types of optimal solutions, namely for minimal, strongly minimal, properly minimal and weakly minimal solutions of a multi objective optimization problem. Finally, an application to vector approximation problems is discussed.

The notation which is used in this article is mostly the same as, for instance, in the book of Holmes[18]. Therefore we recall only some basic definitions. For two subsets S and T of a real linear space and two real numbers α and β the set $\alpha S+\beta T$ is defined as

$$\alpha S+\beta T := \{\alpha s+\beta t \mid s\in S \text{ and } t\in T\}.$$

The *core* of a subset S of a real linear space X is given as

$$\operatorname{cor}(S) := \{s \in S \mid \text{for each } x \in X \text{ there exists some } \bar{\lambda} > 0 \text{ with } s + \lambda x \in S \text{ for all } \lambda \in (0, \bar{\lambda}]\}.$$

A subset S of a real linear space X is called *algebraically closed*, if S equals the set

$$S \cup \{x \in X \mid \text{there exists some } s \in S,\ s \neq x, \text{ with } \lambda s + (1-\lambda)x \in S \text{ for all } \lambda \in (0,1]\},$$

and the set S is said to be *algebraically bounded*, if for each $s \in S$ and each $x \in X$ there exists some $\bar{\lambda} > 0$ such that

$$s + \lambda x \notin S \quad \text{for all } \lambda \geq \bar{\lambda}.$$

The *interior* of a subset S of a real topological linear space is denoted by int(S), and cl(S) denotes the *closure* of S. If X is a real topological linear space, then X' denotes the *algebraic dual space* of X and X^* is used for the *topological dual space* of X. A subset C of a real linear space is called a *cone*, if for each $\lambda \geq 0$ and each $c \in C$ the element λc belongs to C as well. A convex cone C in a real linear space X induces a *partial ordering* $\leq$ on X, if we define for arbitrary $x, y \in X$

$$x \leq y \quad :\Longleftrightarrow \quad y - x \in C.$$

In this case the convex cone C is called the *ordering cone*. A cone C in a real linear space X is called *pointed*, if $(-C) \cap C = \{0_X\}$ where 0_X denotes the zero element in X. In a partially ordered real linear space X with an ordering cone C for arbitrary elements $x, y \in X$ with $y \in \{x\} + C$ the *order interval* between x and y is denoted by

$$[x, y] := (\{x\} + C) \cap (\{y\} - C).$$

If C_X is a convex cone in a real linear space X, then the convex cone

$$C_{X'} := \{x' \in X' \mid x'(x) \geq 0 \text{ for all } x \in C_X\}$$

is called the *dual cone* for C_X, and the set

$$C^{*}_{X'} := \{x'\in X' \mid x'(x)>0 \text{ for all } x\in C_X\setminus\{0_X\}\}$$

is called the *quasi-interior* of the dual cone for C_X.

1.1 Problem Formulation

In this paper we investigate general multi objective optimization problems in real linear spaces. The standard assumption reads as follows:

> Let S be a nonempty subset of a real linear space X, and let Y be a partially ordered real linear space with an ordering cone C_Y. Moreover, let a mapping $f:S\to Y$ be given. (1.1)

Under this assumption we consider the multi objective optimization problem

$$\min_{x\in S} f(x) \, , \tag{1.2}$$

that is , we ask for the "minima" of f on S. The precise definition of these "optima" is given in the next subsection. In problem (1.2) f is called the *objective mapping* and S describes the *constraint set*.

Many problems in the applied sciences can be formulated as problem (1.2). For instance, Stadler[35] gives an overview on complicated problems in mechanical engineering; Kitagawa/Watanabe/Nishimura/Matsubara[26] present some complex problems arising in chemical engineering; Farwick[11], Koller/Farwick[27] and Baier[6] consider problems in structural engineering and Grauer/Lewandowski/Schrattenholzer[16] investigate a problem in environmental engineering.

Example 1.1:

As a simple example consider the design of a beam with a rectangular cross-section and a given length ℓ (see Fig. 1).

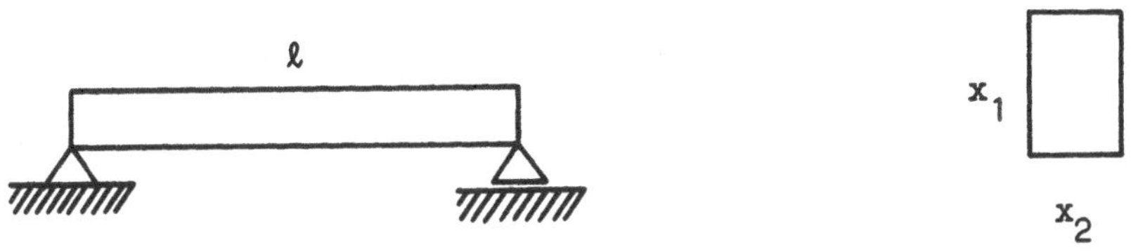

Fig. 1

We ask for an "optimal" height x_1 and width x_2 where the following restrictions should be satisfied (stress condition, stability conditions, nonnegativity conditions):

$$x_1^2 x_2 \geq 2000, \; x_1 \leq 4x_2, \; x_1 \geq x_2, \; x_1 \geq 0, \; x_2 \geq 0.$$

Among all feasible values for x_1 and x_2 we are interested in those which lead to a light and cheap construction. As a criterion for the weight we take the volume $\ell x_1 x_2$ of the beam (where we assume that the material is homogeneous), and as a criterion for the costs we choose the area $\frac{\pi}{4}(x_1^2 + x_2^2)$ of the cross-section of a trunk from which an appropriate beam with the height x_1 and the width x_2 can be cut out. In this special case the assumption (1.1) reduces to $X := \mathbb{R}^2$, $Y := \mathbb{R}^2$, $C_Y := \mathbb{R}^2_+$, $S := \{(x_1, x_2) \in \mathbb{R}^2 \mid x_1^2 x_2 \geq 2000, x_1 \leq 4x_2, x_1 \geq x_2, x_1 \geq 0, x_2 \geq 0\}$ and $f: S \to Y$ is given by

$$f(x_1, x_2) = \begin{pmatrix} \ell x_1 x_2 \\ \frac{\pi}{4}(x_1^2 + x_2^2) \end{pmatrix} \quad \text{for all } (x_1, x_2) \in \mathbb{R}^2 .$$

Before we present a multi objective optimization problem in an infinite dimensional setting we recall the definition of a vectorial norm introduced by Kantorovitch[25].

Definition 1.2:

Let X be a real linear space, and let Y be a partially ordered real linear space with an ordering cone C_Y. A mapping $|||\cdot|||: X \to Y$ is called a ***vectorial norm***, if

a) $|||x||| \in C_Y$ for all $x \in X$; $|||x||| = 0_Y \iff x = 0_X$;

b) $|||\lambda x||| = |\lambda| \; |||x|||$ for all $\lambda \in \mathbb{R}$ and all $x \in X$;

c) $|||x||| + |||z||| - |||x+z||| \in C_Y$ for all $x, z \in X$.

Example 1.3:

Let S be a nonempty subset of a real linear space X, let Y be a partially ordered real linear space with an ordering cone C_Y, let $|||\cdot|||$ be a vectorial norm, and let some $\hat{x} \in X$ be given. The problem of finding an element in S which has a minimal distance to $\hat{x}$ with respect to this vectorial norm is also called a *vector approximation problem*. It can be formalized as

$$\min_{x \in S} |||x - \hat{x}||| .$$

This kind of problems is studied in detail in section 5.

1.2 Optimality Notions

As pointed out before there are different possibilities in order to define "optimal" solutions of the multi objective optimization problem

(1.2). We restrict ourselves to the minimality, strong minimality, proper minimality and weak minimality notion. First, we need the known definition of a tangent cone (e.g., compare Ref. 28 for a normed setting).

Definition 1.4:

Let S be a nonempty subset of a real separated topological linear space X, and let some $\bar{s}\in S$ be given. The *sequential (Bouligand) tangent cone* $T(S,\bar{s})$ to the set S at $\bar{s}$ is defined as

$$T(S,\bar{s}) := \{ \lim_{n\to\infty} \lambda_n (s_n - \bar{s}) \mid \lambda_n \geq 0 \text{ and } s_n \in S \text{ for all } n\in\mathbb{N} \text{ and } \bar{s} = \lim_{n\to\infty} s_n \}.$$

The sequential tangent cone $T(S,\bar{s})$ is a local approximation of the set $S-\{\bar{s}\}$ at 0_X. If $(X, \|\cdot\|_X)$ is a real normed space, then $T(S,\bar{s})$ is closed, and it is even convex, if the set S is convex.

The optimality notions which are investigated in this paper are defined as follows.

Definition 1.5:

Let the assumption (1.1) be satisfied.

a) An element $\bar{x}\in S$ is called a *minimal* point of f on S, if
$(\{f(\bar{x})\}-C_Y)\cap f(S) = \{f(\bar{x})\}$.

b) An element $\bar{x}\in S$ is called a *strongly minimal* point of f on S, if
$f(S) \subset \{f(\bar{x})\}+C_Y$.

c) If, in addition, Y is a real separated topological linear space, then an element $\bar{x}\in S$ is called a *properly minimal* point of f on S, if $\bar{x}$ is a minimal point of f on S and if

$$(-C_Y) \cap \mathrm{cl}(T(f(S)+C_Y, f(\bar{x}))) = \{0_Y\}$$

(i.e., there is no $y \in \mathrm{cl}(T(f(S)+C_Y, f(\bar{x})))$ with $y \in -C_Y \setminus \{0_Y\}$).

d) If, in addition, $\mathrm{cor}(C_Y) \neq \emptyset$, then an element $\bar{x} \in S$ is called a ***weakly minimal*** point of f on S, if $(\{f(\bar{x})\} - \mathrm{cor}(C_Y)) \cap f(S) = \emptyset$.

For the definition of properly minimal points we follow Borwein[7] (in this paper one can also find some relations with the notion of Geoffrion[15]). Obviously, each properly minimal point of f on S is also a minimal point of f on S. If the ordering cone is pointed, then each strongly minimal point is also a minimal point. If $C_Y \neq Y$ and $\mathrm{cor}(C_Y) \neq \emptyset$, then each minimal point of f on S is a weakly minimal point as well (for instance, see Ref. 30 in the case of $Y = \mathbb{R}^n$ and $C_Y = \mathbb{R}^n_+$). As it may be seen from the following example the converse implications are not true in general.

Example 1.6:

Consider $X := Y := \mathbb{R}^2$, $C_Y := \mathbb{R}^2_+$, f=identity and the set
$S := \{(x_1, x_2) \in [0,2] \times [0,2] \mid x_2 \geq 1 - \sqrt{1-(x_1-1)^2} \text{ for all } x_1 \in [0,1]\}$ (see Fig. 2).

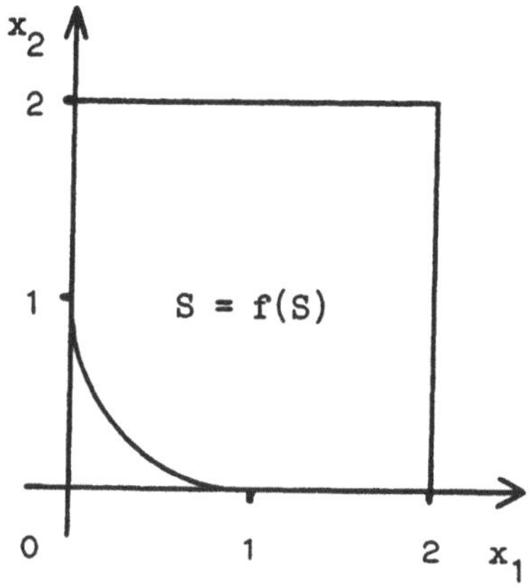

Fig. 2

It is evident that the set M of all minimal points of f on S is given by

$$M = \{(x_1, 1-\sqrt{1-(x_1-1)^2}) \mid x_1 \in [0,1]\}.$$

There exist no strongly minimal points of f on S. The set M_p of all properly minimal points of f on S reads

$$M_p = M \setminus \{(0,1), (1,0)\},$$

and the set M_w of all weakly minimal points of f on S is

$$M_w = M \cup \{(x_1,0) \mid x_1 \in (1,2]\} \cup \{(0,x_2) \mid x_2 \in (1,2]\}.$$

So, for this example we have $M_p \subset M \subset M_w$ and each inclusion is even strict.

2. MONOTONICITY AND GENERALIZED CONVEXITY

For the presentation of necessary and sufficient conditions for optimal solutions of a multi objective optimization problem we need several monotonicity concepts and a generalized convexity notion for the objective mapping. These concepts are introduced in the following subsections.

2.1 Monotonicity Concepts

The following monotonicity concepts are very helpful for our characterization results in the sections 3 and 4.

Definition 2.1:

Let S be a nonempty subset of a subset T of a partially ordered real linear space X with an ordering cone C.

a) A functional $f:T \to \mathbb{R}$ is called *monotonically increasing* on S, if for each $\bar{x} \in S$

$$x \in (\{\bar{x}\}-C) \cap S \quad \Rightarrow \quad f(x) \le f(\bar{x}) .$$

b) A functional $f:T \to \mathbb{R}$ is called *strongly monotonically increasing* on S, if for each $\bar{x} \in S$

$$x \in (\{\bar{x}\}-C) \cap S,\ x \ne \bar{x} \quad \Rightarrow \quad f(x) < f(\bar{x}) .$$

c) If, in addition, $\mathrm{cor}(C) \ne \emptyset$, then a functional $f:T \to \mathbb{R}$ is called *strictly monotonically increasing* on S, if for each $\bar{x} \in S$

$$x \in (\{\bar{x}\}-\mathrm{cor}(C)) \cap S \quad \Rightarrow \quad f(x) < f(\bar{x}) .$$

If $\mathrm{cor}(C) \ne \emptyset$, then each functional which is strongly monotonically increasing on S is also strictly monotonically increasing on S.

Example 2.2:

a) Let S be a nonempty subset of a partially ordered real linear space X with an ordering cone C_X. Each linear functional $\ell \in C_{X'}$ is monotonically increasing on S. Furthermore, each linear functional $\ell \in C^{\#}_{X'}$ is strongly monotonically increasing on S. If $\mathrm{cor}(C_X) \ne \emptyset$, then each linear functional $\ell \in C_{X'} \setminus \{0_{X'}\}$ is strictly monotonically increasing on S.

b) If $X = L_p(\Omega, \mathcal{B}, \mu)$ where $(\Omega, \mathcal{B}, \mu)$ is a measure space and $p \in [1, \infty)$, and if X is equipped with the natural ordering cone

$$C := \{x \in X \mid x(t) \ge 0 \text{ almost everywhere on } \Omega\},$$

then the $L_p(\Omega, \mathcal{B}, \mu)$-norm is strongly monotonically increasing on C.

c) If $X = L_\infty(\Omega, \mathcal{B}, \mu)$ where $(\Omega, \mathcal{B}, \mu)$ is a measure space, then the $L_\infty(\Omega, \mathcal{B}, \mu)$-norm is strictly monotonically increasing on the natural ordering cone

$$C := \{x \in X \mid x(t) \ge 0 \text{ almost everywhere on } \Omega\}.$$

For the proof of this assertion notice that

$$\text{int}(C) = \{x\in X \mid \text{there exists some } \alpha>0 \text{ with } x(t)\geq\alpha \text{ almost everywhere on } \Omega\} \neq \emptyset .$$

Then int(C) equals the core of C, and for each $y\in C$ and each $x\in(\{y\}-\text{cor}(C))\cap C$ it follows $y-x\in\text{cor}(C)$ which implies that there exists some $\alpha>0$ with

$$y(t)-x(t)\geq\alpha \text{ almost everywhere on } \Omega.$$

Consequently, we get

$$\alpha + \operatorname*{ess\,sup}_{t\in\Omega}\{x(t)\} \leq \operatorname*{ess\,sup}_{t\in\Omega}\{y(t)\}$$

and $\|x\|_{L_\infty(\Omega,\mathcal{B},\mu)} < \|y\|_{L_\infty(\Omega,\mathcal{B},\mu)}$.

d) If $X=C(M)$ (real linear space of continuous real-valued functions on M) where M is a compact metric space, then the maximum-norm $\|\cdot\|_{C(M)}$ defined by

$$\|x\|_{C(M)} := \max_{t\in M}\{|x(t)|\} \quad \text{for all } x\in C(M)$$

is strictly monotonically increasing on the natural ordering cone

$$C := \{x\in X \mid x(t)\geq 0 \text{ for all } t\in M\}.$$

The proof of this assertion is similar to that under part c).

e) Let $(X, \langle .,. \rangle)$ be a real Hilbert space with a convex cone C. Then the norm on X is strongly monotonically increasing on C if and only if

$$\langle x,y\rangle \geq 0 \quad \text{for all } x,y\in C. \tag{2.1}$$

This assertion was proved by Rolewicz[33] (the condition (2.1) was given by Wierzbicki[40]).

2.2 Convex-Like Mappings

In multi objective optimization one needs often the assumption that the set $f(S)+C_Y$ (where f, S and C_Y are given as in (1.1)) is convex. Convex mappings fulfill this assumption, but the class of mappings with this property is even larger.

Definition 2.3:

Let the assumption (1.1) be satisfied. The mapping f is called *convex-like*, if for each $x,z \in S$ and each $\lambda \in [0,1]$ there exists some $s \in S$ with

$$\lambda f(x) + (1-\lambda) f(z) - f(s) \in C_Y.$$

Convex-like mappings were introduced by Vogel[37].

Example 2.4:

Let the mapping $f:[\pi,\infty) \to \mathbb{R}^2$ be given by

$$f(x) = \begin{pmatrix} x \\ \sin x \end{pmatrix} \quad \text{for all } x \in [\pi,\infty)$$

where $\mathbb{R}^2$ is partially ordered in a componentwise sense. The mapping f is convex-like (but it is not convex).

Convex-like mappings can be characterized in the following way (for a proof see Ref. 37, p. 165).

Theorem 2.5 (Vogel[37]):

Let the assumption (1.1) be satisfied. Then the mapping f is convex-like if and only if the set $f(S)+C_Y$ is convex.

3. NECESSARY CONDITIONS FOR OPTIMAL POINTS

In this section various necessary conditions for minimal, strongly minimal, properly minimal and weakly minimal points are presented. We begin our discussion with the minimality notion.

<u>Theorem 3.1:</u>

Let the assumption (1.1) be satisfied and, in addition, let the ordering cone C_Y be pointed and algebraically closed and let it have a nonempty core. If $\bar{x}\in S$ is a minimal point of f on S, then for each $\hat{y}\in\{f(\bar{x})\}-\mathrm{cor}(C_Y)$ there exists a norm $\|.\|$ on Y which is monotonically increasing on C_Y with the property

$$1 = \|f(\bar{x})-\hat{y}\| < \|f(x)-\hat{y}\| \quad \text{for all } x\in S \text{ with } f(x)\neq f(\bar{x}).$$

<u>Proof:</u>

For an arbitrary element $\hat{y}\in\{f(\bar{x})\}-\mathrm{cor}(C_Y)$ consider the order interval $[\hat{y}-f(\bar{x}),f(\bar{x})-\hat{y}]$ which is absolutely convex, algebraically bounded and algebraically closed and for which $0_Y\in\mathrm{cor}([\hat{y}-f(\bar{x}),f(\bar{x})-\hat{y}])$ (for further details see Ref. 23). But then the Minkowski functional $\|.\|$ on Y given by

$$\|y\| = \inf_{\lambda>0}\{\lambda \mid \tfrac{1}{\lambda}y\in[\hat{y}-f(\bar{x}),f(\bar{x})-\hat{y}]\} \quad \text{for all } y\in Y$$

is indeed a norm and

$$[\hat{y}-f(\bar{x}),f(\bar{x})-\hat{y}] = \{y\in Y \mid \|y\|\leq 1\}.$$

Since $\bar{x}$ is a minimal point of f on S, we conclude

$$[\hat{y}-f(\bar{x}),f(\bar{x})-\hat{y}]\cap(f(S)-\{\hat{y}\}) = \{f(\bar{x})-\hat{y}\}$$

which implies

$$1 = \|f(\bar{x})-\hat{y}\| < \|f(x)-\hat{y}\| \quad \text{for all } x\in S \text{ with } f(x)\neq f(\bar{x}).$$

In order to show the monotonicity of the norm $\|.\|$ on C_Y take any $y \in C_Y\setminus\{0_Y\}$. Then we get

$$\frac{1}{\|y\|}[0_Y,y] = [0_Y, \frac{1}{\|y\|}y] \subset \{z\in Y \mid \|z\|\leq 1\}$$

and

$$[0_Y,y] \subset \{z\in Y \mid \|z\|\leq\|y\|\}.$$

The last inclusion is even true for $y=0_Y$. This completes the proof. □

The preceding theorem states under suitable assumptions that each minimal point of f on S is a solution of an appropriate approximation problem. So, multi objective optimization problems lead to approximation problems. Theorem 3.1 (and also Theorem 3.8) extend a result given by Wierzbicki[41, 42, 43] for so-called order preserving and order representing functionals (compare also Vogel[39] for convex problems).

It is important to note that for a special multi objective optimization problem in a finite dimensional setting the norm in Theorem 3.1 is even the weighted maximum-norm (see also Ref. 36).

Theorem 3.2:

Let S be a nonempty subset of a real linear space X, and let $f:S\to\mathbb{R}^n$ be a given mapping where the space $\mathbb{R}^n$ is assumed to be partially ordered in a componentwise sense. If $\bar{x}\in S$ is a minimal point of f on S, then for each $\hat{y}\in\mathbb{R}^n$ with

$$\hat{y}_i < f_i(x) \quad \text{for all } i\in\{1,\ldots,n\} \text{ and all } x\in S \tag{3.1}$$

there exist positive real numbers $t_1,\ldots,t_n$ with

$$\max_{1\le i\le n} \{t_i(f_i(\bar{x})-\hat{y}_i)\} < \max_{1\le i\le n} \{t_i(f_i(x)-\hat{y}_i)\} \quad \text{for all } x\in S \text{ with } f(x)\neq f(\bar{x}).$$

Proof:

For any minimal point $\bar{x}\in S$ of f on S and any $\hat{y}\in\mathbb{R}^n$ with the property (3.1) we define the real numbers

$$t_i := \frac{1}{f_i(\bar{x})-\hat{y}_i} > 0 \quad \text{for all } i\in\{1,\dots,n\}$$

and the weighted maximum-norm $\|\cdot\|$ on $\mathbb{R}^n$ by

$$\|y\| := \max_{1\le i\le n} \{t_i|y_i|\} \quad \text{for all } y\in\mathbb{R}^n.$$

Because of

$$f(\bar{x}) \in \{y\in\mathbb{R}^n \mid \|y-\hat{y}\|\le 1\} \subset \{f(\bar{x})\}-\mathbb{R}^n_+$$

we get

$$\{y\in\mathbb{R}^n \mid \|y-\hat{y}\|\le 1\} \cap f(S) = \{f(\bar{x})\}.$$

This set equation leads immediately to the assertion. □

A simpler necessary condition can be obtained, if the objective mapping f is convex-like.

Theorem 3.3:

Let the assumption (1.1) be satisfied and, in addition, let the ordering cone C_Y be pointed and nontrivial (i.e. $C_Y\neq\{0_Y\}$). If f is convex-like and $\mathrm{cor}(f(S)+C_Y)\neq\emptyset$, then for each minimal point $\bar{x}\in S$ of f on S there exists a linear functional $\ell\in C_{Y'}\setminus\{0_{Y'}\}$ with the property

$$(\ell\circ f)(\bar{x}) \le (\ell\circ f)(x) \quad \text{for all } x\in S.$$

Proof:

$\bar{x}\in S$ is assumed to be a minimal point of f on S, i.e.

$$(\{f(\bar{x})\}-C_Y)\cap f(S) = \{f(\bar{x})\}$$

which implies

$$(\{f(\bar{x})\}-C_Y)\cap (f(S)+C_Y) = \{f(\bar{x})\}.$$

Since f is convex-like, by Theorem 2.5 the set $f(S)+C_Y$ is convex and by assumption it has even a nonempty core. Then with a known separation theorem (e.g., see Ref. 18, Cor. 4.B) there exist a linear functional $\ell\in Y'\setminus\{0_{Y'}\}$ and a real number α with

$$\ell(f(\bar{x})-c_1)\le\alpha\le\ell(f(x)+c_2) \quad \text{for all } x\in S \text{ and all } c_1,c_2\in C_Y.$$

Since C_Y is a cone, we obtain immediately

$$\ell(y)\ge 0 \quad \text{for all } y\in C_Y$$

and $\ell\in C_{Y'}\setminus\{0_{Y'}\}$, respectively. Moreover, we get with $c_1=c_2=0_Y$

$$(\ell\circ f)(\bar{x})\le(\ell\circ f)(x) \quad \text{for all } x\in S.$$ □

The result of Theorem 3.3 can also be interpreted in the following way: Under the stated assumptions for each minimal point $\bar{x}$ of f on S there exists a linear functional $\ell\in C_{Y'}\setminus\{0_{Y'}\}$ such that $\bar{x}$ is a solution of the single objective optimization problem

$$\min_{x\in S}\ (\ell\circ f)(x).$$

With the following theorem we formulate a necessary and sufficient condition with linear functionals but without the convex-likeness assumption of Theorem 3.3.

Theorem 3.4:

Let the assumption (1.1) be satisfied and, in addition, let Y be locally convex and let the ordering cone C_Y be closed. An element $\bar{x}\in S$ is a minimal point of f on S if and only if for each $x\in S$ with $f(x)\neq f(\bar{x})$ there exists a continuous linear functional $\ell\in C_Y{}^*\backslash\{0_Y{}^*\}$ with $(\ell\circ f)(\bar{x})<(\ell\circ f)(x)$.

Proof:

Let $\bar{x}\in S$ be a minimal point of f on S, i.e.

$$(\{f(\bar{x})\}-C_Y)\cap f(S) = \{f(\bar{x})\}.$$

This set equation can also be interpreted in the following way:

$$f(x)\notin\{f(\bar{x})\}-C_Y \quad \text{for all } x\in S \text{ with } f(x)\neq f(\bar{x}). \tag{3.2}$$

Since C_Y is closed and convex, the set $\{f(\bar{x})\}-C_Y$ is closed and convex as well, and with a known separation theorem (compare Ref. 18, Cor. 11.F) the statement (3.2) is equivalent to: For each $x\in S$ with $f(x)\neq f(\bar{x})$ there exists a continuous linear functional $\ell\in C_Y{}^*\backslash\{0_Y{}^*\}$ with $(\ell\circ f)(\bar{x})<(\ell\circ f)(x)$. □

Roughly speaking, by Theorem 3.4 $\bar{x}$ is a minimal point of f on S if and only if $C_Y{}^*\backslash\{0_Y{}^*\}$ separates $f(\bar{x})$ from each other element in $f(S)$. Theorem 3.4 is actually not a real scalarization result. But with the same arguments we get a scalarization result for strongly minimal points of f on S which is similar to that of Theorem 3.4.

Theorem 3.5:

Let the assumption (1.1) be satisfied and, in addition, let Y be locally convex and let the ordering cone C_Y be closed. An element $\bar{x}\in S$ is a strongly minimal point of f on S if and only if for each $\ell\in C_Y{}^*$

$$(\ell \circ f)(\bar{x}) \leq (\ell \circ f)(x) \quad \text{for all } x \in S.$$

Proof:

Let $\bar{x} \in S$ be a strongly minimal point of f on S, i.e.

$$f(S) \subset \{f(\bar{x})\} + C_Y . \tag{3.3}$$

Since C_Y is a closed convex cone and Y is locally convex, we conclude

$$C_Y = \{y \in Y \mid \ell(y) \geq 0 \text{ for all } \ell \in C_Y{*}\}$$

(for instance, compare Ref. 38, p. 49). So, the inclusion (3.3) is equivalent to

$$f(S) - \{f(\bar{x})\} \subset \{y \in Y \mid \ell(y) \geq 0 \text{ for all } \ell \in C_Y{*}\}$$

which can also be interpreted in the following way: For each $x \in S$ it follows

$$(\ell \circ f)(\bar{x}) \leq (\ell \circ f)(x) \quad \text{for all } \ell \in C_Y{*} .$$

This leads to the assertion. □

Notice that we do not need any convex-likeness assumption in Theorem 3.5. Thus, a strongly minimal point of f on S is a solution for a whole class of single objective optimization problems. This shows that this optimality notion is indeed very strong.

Next, we turn our attention to the notion of proper minimality.

Theorem 3.6:

Let the assumption (1.1) be satisfied and, in addition, let $(Y, \|.\|_Y)$ be a real normed space and let the ordering cone C_Y have a weakly compact base. For some $\bar{x} \in S$ let the cone generated by the set $T(f(S)+C_Y, f(\bar{x})) \cup (f(S)-\{f(\bar{x})\})$ be weakly closed. If $\bar{x}$ is a properly minimal point of f on S, then for each $\hat{y} \in \{f(\bar{x})\}-C_Y$ with $\hat{y} \neq f(\bar{x})$ there exists an (additional) continuous norm $\|\cdot\|$ on Y which is strongly monotonically increasing on C_Y and which has the property

$$1 = \|f(\bar{x})-\hat{y}\| < \|f(x)-\hat{y}\| \quad \text{for all } x\in S \text{ with } f(x)\neq f(\bar{x}).$$

The proof of this theorem which is rather technical and extensive may be found in Ref. 24. If the objective mapping f is convex-like, a scalarization result which uses certain linear functionals is more interesting.

Theorem 3.7 (Vogel[38], Borwein[7]):

Let the assumption (1.1) be satisfied and, in addition, let Y be a real separated locally convex space where the topology gives Y as the topological dual space of Y^*, and let the ordering cone C_Y be closed with $\text{int}(C_{Y^*})\neq\emptyset$. If f is convex-like, then for each properly minimal point $\bar{x}\in S$ of f on S there exists a continuous linear functional $\ell\in C_{Y^*}^*$ with the property

$$(\ell\circ f)(\bar{x}) \le (\ell\circ f)(x) \quad \text{for all } x\in S.$$

The preceding theorem is comparable with Theorem 3.3 and Theorem 3.5. But now the linear functional ℓ belongs to the quasi-interior of the dual cone.

Finally, we present two necessary conditions for weakly minimal points.

Theorem 3.8:

Let the assumption (1.1) be satisfied and, in addition, let the ordering cone $C_Y\neq Y$ be algebraically closed and let it have a nonempty core. If $\bar{x}\in S$ is a weakly minimal point of f on S, then for each $\hat{y}\in\{f(\bar{x})\}-\text{cor}(C_Y)$

there exists a seminorm $\|.\|$ on Y which is strictly monotonically increasing on $\mathrm{cor}(C_Y)$ with the property

$$1 = \|f(\bar{x})-\hat{y}\| \le \|f(x)-\hat{y}\| \quad \text{for all } x\in S.$$

Proof:

Let $\bar{x}\in S$ be a weakly minimal point of f on S. Take any element $\hat{y}\in\{f(\bar{x})\}-\mathrm{cor}(C_Y)$ and consider the order interval $[\hat{y}-f(\bar{x}),f(\bar{x})-\hat{y}]$. Then a seminorm $\|.\|$ on Y is defined by the Minkowski functional

$$\|y\| = \inf_{\lambda>0} \{\lambda \mid \tfrac{1}{\lambda}y\in[\hat{y}-f(\bar{x}),f(\bar{x})-\hat{y}]\} \quad \text{for all } y\in Y,$$

and we have

$$\mathrm{cor}([\hat{y}-f(\bar{x}),f(\bar{x})-\hat{y}]) = \{y\in Y \mid \|y\|<1\}.$$

Since $\bar{x}$ is a weakly minimal point of f on S, we conclude

$$\{y\in Y \mid \|y\|<1\}\cap(f(S)-\{\hat{y}\}) = \emptyset$$

which implies

$$1\le\|f(x)-\hat{y}\| \quad \text{for all } x\in S.$$

$f(\bar{x})-\hat{y}$ belongs to the algebraic boundary of the order interval $[\hat{y}-f(\bar{x}),f(\bar{x})-\hat{y}]$, and therefore $\|f(\bar{x})-\hat{y}\|=1$. Finally, we show that the seminorm $\|.\|$ is strictly monotonically increasing on $\mathrm{cor}(C_Y)$. For each $y\in\mathrm{cor}(C_Y)$ we get $\|y\|>0$. For the proof of this assertion assume that $\|y\|=0$. Then we obtain

$$\tfrac{1}{\lambda}y\in\{f(\bar{x})-\hat{y}\}-C_Y \quad \text{for all } \lambda>0$$

and

$$-y+\lambda(f(\bar{x})-\hat{y})\in C_Y \quad \text{for all } \lambda>0,$$

respectively. C_Y is algebraically closed, and therefore we conclude $-y\in C_Y$. Since $y\in\mathrm{cor}(C_Y)$ and $-y\in C_Y$, 0_Y belongs to $\mathrm{cor}(C_Y)$ which contradicts the assumption $C_Y \neq Y$. So, for each $y \in \text{cor}(C_Y)$ it follows $\|y\|>0$, and then we get

$$\frac{1}{\|y\|}\,\text{cor}([0_Y,y]) = \text{cor}([0_Y, \frac{1}{\|y\|}y]) \subset \{z\in Y \mid \|z\|<1\}$$

and

$$\text{cor}([0_Y,y]) \subset \{z\in Y \mid \|z\| < \|y\|\}.$$

This inclusion implies that the seminorm $\|.\|$ is strictly monotonically increasing on $\text{cor}(C_Y)$. □

It can be expected that for the weak minimality notion a special scalarization result can be formulated under a convex-likeness assumption as well. This is done in

Theorem 3.9:

Let the assumption (1.1) be satisfied and, in addition, let the ordering cone C_Y have a nonempty core. If f is convex-like, then for each weakly minimal point $\bar{x}\in S$ of f on S there exists a linear functional $\ell \in C_{Y'}\setminus\{0_{Y'}\}$ with the property

$$(\ell\circ f)(\bar{x}) \leq (\ell\circ f)(x) \quad \text{for all } x\in S.$$

Proof:

Let $\bar{x}\in S$ be a weakly minimal point of f on S. Then it follows

$$(\{f(\bar{x})\}-\text{cor}(C_Y)) \cap (f(S)+C_Y) = \emptyset.$$

By Theorem 2.5 the set $f(S)+C_Y$ is convex. Then by a known separation theorem (for instance, compare Ref. 18, Cor. 4.B) there exist a linear functional $\ell\in Y'\setminus\{0_{Y'}\}$ and a real number α with

$$\ell(f(\bar{x})-c_1) \leq \alpha \leq \ell(f(x)+c_2) \quad \text{for all } x\in S \text{ and all } c_1,c_2\in C_Y.$$

Since C_Y is a cone, we get $\ell\in C_{Y'}\setminus\{0_{Y'}\}$ and the assertion is obvious. □

In this section we presented mainly two types of scalarization results, namely a general version via approximation problems (Theorem 3.1, Theorem 3.6 and Theorem 3.8) and a version for convex-like objective mappings with the aid of linear functionals (Theorem 3.3, Theorem 3.7 and Theorem 3.9). Only Theorem 3.4 and Theorem 3.5 are scalarization results with linear functionals without assuming the objective mapping to be convex-like.

4. SUFFICIENT CONDITIONS FOR OPTIMAL POINTS

It is the aim of this section to investigate under which assumptions the necessary conditions presented in the preceding section are also sufficient for optimal points. We begin our discussion with the minimality notion.

<u>Lemma 4.1:</u>

Let the assumption (1.1) be satisfied, let $g:f(S)\to\mathbb{R}$ be a given functional, and let some $\bar{x}\in S$ be given with the property

$$(g\circ f)(\bar{x}) \le (g\circ f)(x) \quad \text{for all } x\in S. \tag{4.1}$$

a) If the functional g is monotonically increasing on f(S) and if strict inequality holds in (4.1) for all $x\in S$ with $f(x)\ne f(\bar{x})$, then $\bar{x}$ is a minimal point of f on S.

b) If the functional g is strongly monotonically increasing on f(S), then $\bar{x}$ is a minimal point of f on S.

Proof:

For the proof of both parts we assume that $\bar{x}$ is not a minimal point of f on S. Then there exists an element $x \in S$ with $f(x) \in (\{f(\bar{x})\}-C_Y)$ and $f(x) \neq f(\bar{x})$. This implies $(g \circ f)(x) \leq (g \circ f)(\bar{x})$ in part a) which contradicts the assumption that strict inequality holds in (4.1) for x, i.e. $(g \circ f)(\bar{x}) < (g \circ f)(x)$. In part b) we obtain $(g \circ f)(x) < (g \circ f)(\bar{x})$ which is a contradiction to the minimality of $g \circ f$ on S. □

Next, we apply Lemma 4.1 to special classes of functionals g, namely to certain seminorms and linear functionals.

Theorem 4.2:

Let the assumption (1.1) be satisfied. Moreover, let $\|.\|$ be a seminorm on Y, and let an element $\hat{y} \in Y$ with

$$f(S) \subset \{\hat{y}\}+C_Y \tag{4.2}$$

and an element $\bar{x} \in S$ with

$$\|f(\bar{x})-\hat{y}\| \leq \|f(x)-\hat{y}\| \quad \text{for all } x \in S \tag{4.3}$$

be given.

a) If the seminorm $\|.\|$ is monotonically increasing on C_Y and if strict inequality holds in (4.3) for all $x \in S$ with $f(x) \neq f(\bar{x})$, then $\bar{x}$ is a minimal point of f on S.

b) If the seminorm $\|.\|$ is strongly monotonically increasing on C_Y, then $\bar{x}$ is a minimal point of f on S.

Proof:

We proof only part a) of the assertion. The proof of the other part is similar. In order to be able to apply Lemma 4.1, a) we show the monotonicity of the functional $\|\cdot-\hat{y}\|$ on $f(S)$. For each $\bar{y}\in f(S)$ we obtain with the inclusion (4.2)

$$(\{\bar{y}\}-C_Y)\cap f(S) \subset (\{\bar{y}\}-C_Y)\cap(\{\hat{y}\}+C_Y) = [\hat{y},\bar{y}] = \{\hat{y}\} + [0_Y,\bar{y}-\hat{y}] .$$

Consequently, we have for each $y\in(\{\bar{y}\}-C_Y)\cap f(S)$

$$y-\hat{y}\in[0_Y,\bar{y}-\hat{y}] .$$

So, we conclude because of the monotonicity of the seminorm $\|\cdot\|$ on C_Y

$$\|y-\hat{y}\| \le \|\bar{y}-\hat{y}\| .$$

Hence, the functional $\|\cdot-\hat{y}\|$ is monotonically increasing on $f(S)$. This completes the proof. □

In Example 2.2, b), e) we discussed already in which normed spaces the norm is strongly monotonically increasing on the ordering cone. So, Theorem 4.2, b) is directly applicable in these cases.

Theorem 4.2, a) generalizes also corresponding results of Rolewicz[33] and Vogel[39]. For $Y=\mathbb{R}^n$ and $C_Y=\mathbb{R}^n_+$ approximation problems of the above kind were also investigated, among others, by Dinkelbach[9], Salukvadze[34], Dinkelbach/Dürr[10], Huang[19], Yu[44], Yu/Leitmann[45] and Gearhart[13].

Theorem 3.1 and Theorem 4.2, a) lead to a characterization of minimal points.

Corollary 4.3:

Let the assumption (1.1) be satisfied and, in addition, let the ordering cone C_Y be pointed and algebraically closed and let it have a nonempty core. Moreover, let an element $\hat{y}\in Y$ with $f(S)\subset\{\hat{y}\}+\mathrm{cor}(C_Y)$ be given. An element $\bar{x}\in S$ is a minimal point of f on S if and only if there exists a norm $\|.\|$ on Y which is monotonically increasing on C_Y with the property

$$\|f(\bar{x})-\hat{y}\| < \|f(x)-\hat{y}\| \quad \text{for all } x\in S \text{ with } f(x)\neq f(\bar{x}).$$

For a special multi objective optimization problem in a finite dimensional setting (compare Theorem 3.2) we obtain immediately the following characterization of minimal points.

<u>Corollary 4.4:</u>

Let S be a nonempty subset of a real linear space X, and let $f:S\to\mathbb{R}^n$ be a given mapping where the space $\mathbb{R}^n$ is assumed to be partially ordered in a componentwise sense. Moreover, let an element $\hat{y}\in\mathbb{R}^n$ with

$$\hat{y}_i < f_i(x) \quad \text{for all } i\in\{1,\ldots,n\} \text{ and all } x\in S$$

be given. An element $\bar{x}\in S$ is a minimal point of f on S if and only if there exist positive real numbers $t_1,\ldots,t_n$ with the property

$$\max_{1\le i\le n} \{t_i(f_i(\bar{x})-\hat{y}_i)\} < \max_{1\le i\le n} \{t_i(f_i(x)-\hat{y}_i)\} \quad \text{for all } x\in S \text{ with } f(x)\neq f(\bar{x}).$$

The preceding corollary which can essentially be found in a paper of Steuer/Choo[36] is the basis for their so-called interactive weighted Chebyshev method.

If the set f(S) has not a lower bound $\hat{y}$, i.e., the inclusion (4.2) is not fulfilled, nevertheless approximation problems are qualified for the determination of minimal points (this idea is due to Rolewicz[33]).

<u>Theorem 4.5:</u>

Let the assumption (1.1) be satisfied. Moreover, let a seminorm $\|.\|$ on Y and an element $\tilde{x}\in S$ be given such that for some $\bar{x}\in S$ with $f(\bar{x})\in\{f(\tilde{x})\}-C_Y$

$$\|f(\bar{x})-f(\tilde{x})\| \geq \|f(x)-f(\tilde{x})\| \quad \text{for all } x\in S \text{ with } f(x)\in\{f(\tilde{x})\}-C_Y. \quad (4.4)$$

a) If the seminorm $\|.\|$ is monotonically increasing on C_Y and if strict inequality holds in (4.4) for all $x\in S$ with $f(x)\in\{f(\tilde{x})\}-C_Y$ and $f(x)\neq f(\bar{x})$, then $\bar{x}$ is a minimal point of f on S.

b) If the seminorm $\|.\|$ is strongly monotonically increasing on C_Y, then $\bar{x}$ is a minimal point of f on S.

Proof:

We proof only part a) of this theorem. First, we show that the functional $-\|.-f(\tilde{x})\|$ is monotonically increasing on $\{f(\tilde{x})\}-C_Y$. For that purpose we take any arbitrary $\bar{y}\in\{f(\tilde{x})\}-C_Y$ and choose any

$$y \in (\{\bar{y}\}-C_Y)\cap(\{f(\tilde{x})\}-C_Y) = \{\bar{y}\}-C_Y .$$

Then we have $f(\tilde{x})-y\in\{f(\tilde{x})-\bar{y}\}+C_Y$ and $f(\tilde{x})-\bar{y}\in\{f(\tilde{x})-y\}-C_Y$. But we have also $f(\tilde{x})-\bar{y}\in C_Y$. Because of the monotonicity of the seminorm on C_Y we obtain

$$\|f(\tilde{x})-\bar{y}\| \leq \|f(\tilde{x})-y\|$$

and

$$-\|\bar{y}-f(\tilde{x})\| \geq -\|y-f(\tilde{x})\| ,$$

respectively. Then Lemma 4.1, a) is applicable and $\bar{x}$ is a minimal point of f on $\{x\in S \mid f(x)\in\{f(\tilde{x})\}-C_Y\}$, i.e.

$$(\{f(\bar{x})\}-C_Y)\cap f(S)\cap(\{f(\tilde{x})\}-C_Y) = \{f(\bar{x})\} .$$

Finally, the inclusion $(\{f(\bar{x})\}-C_Y)\cap f(S) \subset \{f(\tilde{x})\}-C_Y$ leads to

$$(\{f(\bar{x})\}-C_Y)\cap f(S) = \{f(\bar{x})\} ,$$

i.e., $\bar{x}$ is a minimal point of f on S. □

Notice that in Theorem 4.2 we have to determine a minimal "distance" between $\hat{y}$ and the set $f(S)$ whereas in Theorem 4.5 a maximal "distance" between $f(\tilde{x})$ and elements in the set $f(S) \cap (\{f(\tilde{x})\} - C_Y)$ has to be determined.

Now, we study certain linear functionals.

Theorem 4.6:

Let the assumption (1.1) be satisfied.

a) If there exist a linear functional $\ell \in C_{Y'}$ and an element $\bar{x} \in S$ with

$$(\ell \circ f)(\bar{x}) < (\ell \circ f)(x) \quad \text{for all } x \in S \text{ with } f(x) \neq f(\bar{x}),$$

then $\bar{x}$ is a minimal point of f on S.

b) If there exist a linear functional $\ell \in C^*_{Y'}$ and an element $\bar{x} \in S$ with

$$(\ell \circ f)(\bar{x}) \leq (\ell \circ f)(x) \quad \text{for all } x \in S,$$

then $\bar{x}$ is a minimal point of f on S.

The proof of Theorem 4.6 follows directly from Lemma 4.1 and the remarks in Example 2.2, a). Part b) of the preceding theorem was proved by Hurwicz[20].

Notice that a theorem of Krein-Rutman (Ref. 29, p. 218, and Ref. 8, p. 425) gives conditions under which the set $C^*_{Y^*}$ is nonempty (in a real separable normed space $(Y, \|\cdot\|_Y)$ with a closed pointed convex cone C_Y the set $C^*_{Y^*}$ is nonempty). If we compare Theorem 3.3 and Theorem 4.6 we see that one cannot prove the sufficiency of the necessary condition formulated in Theorem 3.3. So, one cannot present a complete characterization like Corollary 4.3 for linear functionals instead of norms.

Since we characterized already strongly minimal points in Theorem 3.5, we study now the proper minimality notion.

Theorem 4.7:

Let the assumption (1.1) be satisfied and, in addition, let $(Y, \|.\|_Y)$ be a real normed space and let the ordering cone C_Y have a nonempty core. Let $\|.\|$ be any (additional) continuous norm on Y which is strongly monotonically increasing on C_Y. Moreover, let an element $\hat{y}\in Y$ with $f(S) \subset \{\hat{y}\}+\mathrm{cor}(C_Y)$ be given. If there exists an element $\bar{x}\in S$ with the property

$$\|f(\bar{x})-\hat{y}\| \leq \|f(x)-\hat{y}\| \quad \text{for all } x\in S, \tag{4.5}$$

then $\bar{x}$ is a properly minimal point of f on S.

Proof:

Since the norm $\|.\|$ is strongly monotonically increasing on C_Y and $f(S)-\{\hat{y}\} \subset \mathrm{cor}(C_Y)$, by Lemma 4.1, b) $\bar{x}$ is a minimal point of f on S. Next, we proof that

$$(-C_Y) \cap \mathrm{cl}(T(f(S)+C_Y, f(\bar{x}))) = \{0_Y\} .$$

The sequential tangent cone is always closed in a normed setting, and therefore we show only

$$(-C_Y) \cap T(f(S)+C_Y, f(\bar{x})) = \{0_Y\} .$$

Since the norm $\|.\|$ is assumed to be strongly monotonically increasing on C_Y, we obtain from (4.5)

$$\|f(\bar{x})-\hat{y}\| \leq \|f(x)-\hat{y}\| \leq \|f(x)+c-\hat{y}\| \quad \text{for all } x\in S \text{ and all } c\in C_Y$$

and

$$\|f(\bar{x})-\hat{y}\| \leq \|y-\hat{y}\| \quad \text{for all } y \in f(S)+C_Y , \tag{4.6}$$

respectively. It is evident that the functional $\|.-\hat{y}\|$ is convex. But it is also continuous in the topology generated by the norm $\|\cdot\|_Y$. Then by a known result (for instance, see Ref. 28, Thm. III.2.2) the inequality (4.6) implies

$$\|f(\bar{x})-\hat{y}\| \leq \|f(\bar{x})-\hat{y}+h\| \quad \text{for all } h \in T(f(S)+C_Y, f(\bar{x})) . \tag{4.7}$$

With $T := T(f(S)+C_Y, f(\bar{x})) \cap (\{\hat{y}-f(\bar{x})\}+C_Y)$ the inequality (4.7) is also true for all $h \in T$, and by Lemma 4.1, b) we get

$$(-C_Y) \cap T = \{0_Y\} . \tag{4.8}$$

Now, we assume that $(-C_Y) \cap T(f(S)+C_Y, f(\bar{x})) \neq \{0_Y\}$. Then there exists some $y \in (-C_Y) \cap T(f(S)+C_Y, f(\bar{x}))$ with $y \neq 0_Y$. Because of the inclusion $f(S) \subset \{\hat{y}\}+\text{cor}(C_Y)$ there exists some $\lambda>0$ with $\lambda y \in \{\hat{y}-f(\bar{x})\}+C_Y$. Consequently, we get

$$\lambda y \in (-C_Y) \cap T(f(S)+C_Y, f(\bar{x})) \cap (\{\hat{y}-f(\bar{x})\}+C_Y) ,$$

and therefore we have $\lambda y \in (-C_Y) \cap T$ which contradicts the set equation (4.8). Hence, the set equation $(-C_Y) \cap T(f(S)+C_Y, f(\bar{x})) = \{0_Y\}$ is true and the assertion is obvious. □

A similar sufficiency result for properly minimal points can be found in a paper of Vogel[39]. For $Y=\mathbb{R}^n$ and $C_Y=\mathbb{R}^n_+$ a corresponding result was proved by Dinkelbach/Dürr[10].

In Theorem 3.6 we do not need the assumptions $\text{cor}(C_Y) \neq \emptyset$ and $\hat{y} \in \{f(\bar{x})\}-\text{cor}(C_Y)$ which play an important role in Theorem 4.7. On the other hand in Theorem 4.7 it is not required that strict inequality holds in (4.5) for all $x \in S$ with $f(x) \neq f(\bar{x})$. With Theorem 3.6 and Theorem 4.7 we get immediately a characterization of properly minimal points.

Corollary 4.8:

Let the assumption (1.1) be satisfied and, in addition, let $(Y, \|.\|_Y)$ be a real normed space and let the ordering cone C_Y have a nonempty core and a weakly compact base. Moreover, let an element $\hat{y}\in Y$ with $f(S) \subset \{\hat{y}\}+\text{cor}(C_Y)$ be given, and for some $\bar{x}\in S$ let the cone generated by the set $T(f(S)+C_Y,f(\bar{x})) \cup (f(S)-\{f(\bar{x})\})$ be weakly closed. Then $\bar{x}$ is a properly minimal point of f on S if and only if there exists an (additional) continuous norm $\|.\|$ on Y which is strongly monotonically increasing on C_Y and which has the property

$$1 = \|f(\bar{x})-\hat{y}\| < \|f(x)-\hat{y}\| \quad \text{for all } x\in S \text{ with } f(x)\neq f(\bar{x}) .$$

In the preceding corollary we assume that the ordering cone C_Y has a nonempty core and a weakly compact base. Then it can be shown that there exists an (additional) norm $\|.\|$ on Y such that the real normed space $(Y, \|.\|)$ is even reflexive. This shows that the assumptions of Corollary 4.8 are very restrictive.

Another sufficient condition for properly minimal points is given by

Theorem 4.9 (Vogel[38], Borwein[7]):

Let the assumption (1.1) be satisfied and, in addition, let Y be a real separated topological linear space and let $C_{Y^*}^*$ be nonempty. If there exist a continuous linear functional $\ell\in C_{Y^*}^*$ and an element $\bar{x}\in S$ with the property

$$(\ell\circ f)(\bar{x}) \leq (\ell\circ f)(x) \quad \text{for all } x\in S,$$

then $\bar{x}$ is a properly minimal point of f on S.

With Theorem 3.7 and Theorem 4.9 we formulate the following characterization of properly minimal points for convex-like objective mappings.

Corollary 4.10 (Vogel[38], Borwein[7]):

Let the assumption (1.1) be satisfied and, in addition, let Y be a real separated locally convex linear space where the topology gives Y as the topological dual space of Y^*, and let the ordering cone C_Y be closed with $\text{int}(C_{Y^*})\neq\emptyset$. Moreover, let f be convex-like. An element $\bar{x}\in S$ is a properly minimal point of f on S if and only if there exists a continuous linear functional $\ell\in C^{*}_{Y^*}$ with the property

$$(\ell\circ f)(\bar{x}) \leq (\ell\circ f)(x) \quad \text{for all } x\in S.$$

From a theoretical point of view this characterization of properly minimal points is very important. For instance, it can be used for a general duality theory (e.g., see Ref. 21). For $Y=\mathbb{R}^n$ further characterizations of properly minimal points can be found in Refs. 17 and 14.

Finally, we turn our attention to the weak minimality notion. For the following results we need a basic lemma.

Lemma 4.11:

Let the assumption (1.1) be satisfied and, in addition, let the ordering cone C_Y have a nonempty core. Moreover, let $g:f(S)\rightarrow\mathbb{R}$ be a given functional which is strictly monotonically increasing on f(S). If there exists an element $\bar{x}\in S$ with the property

$$(g\circ f)(\bar{x}) \leq (g\circ f)(x) \quad \text{for all } x\in S,$$

then $\bar{x}$ is a weakly minimal point of f on S.

Proof:

If $\bar{x}\in S$ is not a weakly minimal point of f on S, then we have $g(f(x)) < g(f(\bar{x}))$ for some $x\in S$ with $f(x) \in \{f(\bar{x})\}-\mathrm{cor}(C_Y)$ which is a contradiction to the minimality of $g\circ f$ on S. □

Theorem 4.12:

Let the assumption (1.1) be satisfied and, in addition, let the ordering cone C_Y have a nonempty core. Moreover, let $\|.\|$ be a seminorm on Y which is strictly monotonically increasing on C_Y, and let an element $\hat{y}\in Y$ with $f(S) \subset \{\hat{y}\}+C_Y$ be given. If there exists an element $\bar{x}\in S$ with

$$\|f(\bar{x})-\hat{y}\| \leq \|f(x)-\hat{y}\| \quad \text{for all } x\in S,$$

then $\bar{x}$ is a weakly minimal point of f on S.

The proof of this theorem is analogous to the proof of Theorem 4.2. But now we apply Lemma 4.11 instead of Lemma 4.1.

In Example 2.2, c), d) we investigated already certain norms which are strictly monotonically increasing on the ordering cone. For these norms Theorem 4.12 is directly applicable.

With Theorem 3.8 and a slightly different version of Theorem 4.12 we get the following characterization of weakly minimal points.

Corollary 4.13:

Let the assumption (1.1) be satisfied and, in addition, let the ordering cone $C_Y \neq Y$ be algebraically closed and let it have a nonempty core.

Moreover, let an element $\hat{y}\in Y$ with $f(S)\subset\{\hat{y}\}+\mathrm{cor}(C_Y)$ be given. An element $\bar{x}\in S$ is a weakly minimal point of f on S if and only if there exists a seminorm $\|.\|$ on Y which is strictly monotonically increasing on $\mathrm{cor}(C_Y)$ with the property

$$\|f(\bar{x})-\hat{y}\|\leq\|f(x)-\hat{y}\| \quad \text{for all } x\in S. \tag{4.9}$$

In contrast to Corollary 4.3 concerning minimal points we do not require in Corollary 4.13 that strict inequality holds in (4.9) for all $x\in S$ with $f(x)\neq f(\bar{x})$. For the following result we do not need the assumption that a "strict" lower bound $\hat{y}$ exists.

Theorem 4.14:

Let the assumption (1.1) be satisfied and, in addition, let the ordering cone C_Y have a nonempty core. Moreover, let an element $\tilde{x}\in S$ and a seminorm $\|.\|$ on Y be given which is strictly monotonically increasing on C_Y. If there exists an element $\bar{x}\in S$ with $f(\bar{x})\in\{f(\tilde{x})\}-C_Y$ and

$$\|f(\bar{x})-f(\tilde{x})\|\geq\|f(x)-f(\tilde{x})\| \quad \text{for all } x\in S \text{ with } f(x)\in\{f(\tilde{x})\}-C_Y,$$

then $\bar{x}$ is a weakly minimal point of f on S.

The proof of Theorem 4.14 is similar to that of Theorem 4.5. The next theorem is evident with the aid of Lemma 4.11.

Theorem 4.15:

Let the assumption (1.1) be satisfied and, in addition, let the ordering cone C_Y have a nonempty core. If for some $\bar{x}\in S$ there exists a linear functional $\ell\in C_{Y'}\setminus\{0_{Y'}\}$ with the property

$$(\ell\circ f)(\bar{x})\leq(\ell\circ f)(x) \quad \text{for all } x\in S,$$

then $\bar{x}$ is a weakly minimal point of f on S.

Although we cannot formulate a complete characterization of minimal points with the aid of linear functionals (compare Theorem 3.3 and Theorem 4.6), this can be done for weakly minimal points.

Corollary 4.16:

Let the assumption (1.1) be satisfied and, in addition, let the ordering cone C_Y have a nonempty core. Moreover, let f be convex-like. An element $\bar{x}\in S$ is a weakly minimal point of f on S if and only if there exists a linear functional $\ell\in C_{Y'}\setminus\{0_{Y'}\}$ with the property

$$(\ell\circ f)(\bar{x}) \leq (\ell\circ f)(x) \quad \text{for all } x\in S.$$

The preceding corollary follows from Theorem 3.9 and Theorem 4.15.

5. APPLICATION TO VECTOR APPROXIMATION PROBLEMS

In this section we investigate special multi objective optimization problems, namely vector approximation problems which were already introduced in Example 1.3. We have the following basic assumption:

Let S be a nonempty subset of a real linear space X, and let Y be a partially ordered real linear space with an ordering cone C_Y. Moreover, let a vectorial norm $|||\cdot|||: X\to Y$ (compare Definition 1.2) and an element $\hat{x}\in X$ be given. (5.1)

Under this assumption we examine the *vector approximation problem*

$$\min_{x\in S} |||x-\hat{x}||| . \qquad (5.2)$$

Problems of this type are discussed, for instance, in Ref. 22 (see also Refs. 12, 4, 5, 1, 2 and 3).

Example 5.1:

We consider the free boundary Stefan problem (investigated by Reemtsen[32])

$$u_{xx}(x,t) - u_t(x,t) = 0, \ (x,t)\in D(s), \qquad (5.3)$$

$$u_x(0,t) = g(t), \ 0<t\leq T, \qquad (5.4)$$

$$u(s(t),t) = 0, \ 0<t\leq T, \qquad (5.5)$$

$$u_x(s(t),t) = -\dot{s}(t), \ 0<t\leq T, \qquad (5.6)$$

$$s(0) = 0 \qquad (5.7)$$

where $g\in C([0,T])$ (real linear space of continuous real-valued functions on $[0,T]$) is a nonpositive function satisfying $g(0)<0$ and

$$D(s) := \{(x,t)\in\mathbb{R}^2 \mid 0<x<s(t), \ 0<t\leq T\} \quad \text{for } s\in C([0,T]) .$$

For the approximative solution of this problem we choose

$$\bar{u}(x,t,a) = \sum_{i=0}^{\ell} a_i v_i(x,t)$$

with

$$v_i(x,t) = \sum_{k=0}^{[\frac{i}{2}]} \frac{i!}{(i-2k)!k!} x^{i-2k} t^k$$

($[\frac{i}{2}]$ denotes the biggest integer less than or equal to $\frac{i}{2}$) and

$$\bar{s}(t,b) = -g(0)t + \sum_{i=1}^{p} b_i t^{i+1}$$

(compare Ref. 32, p. 32). For each $a\in\mathbb{R}^{\ell+1}$ $\bar{u}$ satisfies the partial differential equation (5.3) and for each $b \in \mathbb{R}^p$ $\bar{s}$ satisfies the equation (5.7). Inserting $\bar{u}$, $\bar{s}$ into (5.4), (5.5) and (5.6) we obtain the error functions $\rho_1, \rho_2, \rho_3 \in C([0,T])$ with

$$\rho_1(t,a,b) := \bar{u}_x(0,t,a) - g(t) = \sum_{\substack{i=1 \\ i \text{ odd}}}^{\ell} a_i \frac{i!}{(\frac{i-1}{2})!} t^{\frac{i-1}{2}} - g(t),$$

$$\rho_2(t,a,b) := \bar{u}(\bar{s}(t,b),t,a) = \sum_{i=0}^{\ell} a_i v_i(\bar{s}(t,b),t)$$

and

$$\rho_3(t,a,b) := \bar{u}_x(\bar{s}(t,b),t,a) + \dot{\bar{s}}(t,b)$$
$$= \sum_{i=1}^{\ell} a_i v_{i_x}(\bar{s}(t,b),t) + \dot{\bar{s}}(t,b) .$$

If $\|.\|$ is any norm on $C([0,T])$, then the mapping $|||.|||: C([0,T])^3 \to \mathbb{R}^3$ given by

$$|||(y_1,y_2,y_3)||| := (\|y_1\|,\|y_2\|,\|y_3\|) \quad \text{for all } y_1,y_2,y_3 \in C([0,T])$$

is a vectorial norm, if the space $\mathbb{R}^3$ is assumed to be partially ordered in a componentwise sense. Then we formulate the following vector approximation problem

$$\min_{(a,b) \in \mathbb{R}^{\ell+1} \times \mathbb{R}^p} (\|\rho_1(.,a,b)\|,\|\rho_2(.,a,b)\|,\|\rho_3(.,a,b)\|) .$$

Mathematically, a multi objective approximation problem can also be treated as a so-called *simultaneous approximation problem*:

$$\min_{x \in S} \| \, |||x-\hat{x}||| \, \| \tag{5.8}$$

where $\|.\|$ is a norm on Y. This is a single objective optimization problem.

With the aid of some characterization results presented in the preceding section we can formulate a relationship between the vector approximation problem (5.2) and the simultaneous approximation problem (5.8).

Theorem 5.2:

Let the assumption (5.1) be satisfied and, in addition, let the ordering cone C_Y be pointed and algebraically closed and let it have a nonempty core. Moreover, assume that

$$|||x-\hat{x}||| \in \mathrm{cor}(C_Y) \quad \text{for all } x \in S. \tag{5.9}$$

An element $\bar{x} \in S$ is a minimal point of $|||.-\hat{x}|||$ on S if and only if there exists a norm $\|.\|$ on Y which is monotonically increasing on C_Y with the property

$$\| \, |||\bar{x}-\hat{x}||| \, \| < \| \, |||x-\hat{x}||| \, \| \quad \text{for all } x \in S \text{ with } |||x-\hat{x}||| \neq |||\bar{x}-\hat{x}||| .$$

Proof:

This theorem follows immediately from Corollary 4.3, since the condition (5.9) is equivalent to

$$\{ |||x-\hat{x}||| \mid x \in S\} \subset \{0_Y\} + \mathrm{cor}(C_Y) . \qquad \Box$$

Theorem 5.3:

Let the assumption (5.1) be satisfied and, in addition, let the ordering cone $C_Y \neq Y$ be algebraically closed and let it have a nonempty core. Moreover, assume that the condition (5.9) is satisfied. An element $\bar{x} \in S$ is a weakly minimal point of $|||.-\hat{x}|||$ on S if and only if there exists a seminorm $\|.\|$ on Y which is strictly monotonically increasing on $\mathrm{cor}(C_Y)$ with the property

$$\| \, ||| \bar{x}-\hat{x} ||| \, \| \leq \| \, ||| x-\hat{x} ||| \, \| \quad \text{for all } x \in S .$$

Proof:

We apply Corollary 4.13. □

The last two theorems show the narrow connection between vector approximation and simultaneous approximation. After all, certain simultaneous approximation problems are scalarized vector approximation problems. Theorem 5.2 and Theorem 5.3 extend some results of Bacopoulos/ Godini/Singer[3] (see also Refs. 4, 5, 1 and 2).

6. CONCLUSION

In this paper some single objective optimization problems are presented whose optimal solutions are minimal, strongly minimal, properly minimal or weakly minimal points of an objective mapping on a constraint set. If the objective mapping is convex-like, certain linear functionals can be used for this scalarization. But in general these functionals are seminorms or even norms. The main results are that under suitable assumptions minimal, properly minimal and weakly minimal points can be characterized as a solution of an appropriate approximation problem. This shows the importance of the approximation theory for the theory of multi objective optimization.

REFERENCES

1. Bacopoulos, A., Godini, G. and Singer, I., On best approximation in vector-valued norms, *Colloquia Mathematica Societatis János Bolyai*, 19, 89, 1978.

2. Bacopoulos, A., Godini, G. and Singer, I., Infima of sets in the plane and applications to vectorial optimization, *Revue Roumaine de Mathématiques Pures et Appliquées*, 23, 343, 1978.

3. Bacopoulos, A., Godini, G. and Singer, I., On infima of sets in the plane and best approximation, simultaneous and vectorial, in a linear space with two norms, in: *Special Topics of Applied Mathematics*, Frehse, J., Pallaschke, D. and Trottenberg, U., Eds., North-Holland, Amsterdam, 1980.

4. Bacopoulos, A. and Singer, I., On convex vectorial optimization in linear spaces, *Journal of Optimization Theory and Applications*, 21, 175, 1977.

5. Bacopoulos, A. and Singer, I., Errata corrige: On vectorial optimization in linear spaces, *Journal of Optimization Theory and Applications*, 23, 473, 1977.

6. Baier, H., Mathematische Programmierung zur Optimierung von Tragwerken insbesondere bei mehrfachen Zielen, Dissertation, TH Darmstadt, 1978.

7. Borwein, J.M., Proper efficient points for maximizations with respect to cones, *SIAM Journal on Control and Optimization*, 15, 57, 1977.

8. Borwein, J.M., Continuity and differentiability properties of convex operators, *Proceedings of the London Mathematical Society*, 44, 420, 1982.

9. Dinkelbach, W., Über einen Lösungsansatz zum Vektormaximumproblem, in: *Unternehmensforschung Heute*, Beckmann, M., Ed.,Springer, Berlin, 1971.

10. Dinkelbach, W. and Dürr, W., Effizienzaussagen bei Ersatzprogrammen zum Vektormaximumproblem, *Operations Research Verfahren*, XII, 69, 1972.

11. Farwick, H., Systematisches Entwickeln, Selektieren und Optimierungen von Konstruktionen, ein Beitrag zur Konstruktionssystematik, Dissertation, TH Aachen, 1974.

12. Gearhart, W.B., On vectorial approximation, *Journal of Approximation Theory*, 10, 49, 1974.

13. Gearhart, W.B., Compromise solutions and estimation of the noninferior set, *Journal of Optimization Theory and Applications*, 28, 29, 1979.

14. Gearhart, W.B., Characterization of properly efficient solutions by generalized scalarization methods, *Journal of Optimization Theory and Applications*, 41, 491, 1983.

15. Geoffrion, A.M., Proper efficiency and the theory of vector maximization, *Journal of Mathematical Analysis and Applications*, 22, 618, 1968.

16. Grauer, M., Lewandowski, A. and Schrattenholzer, L., Use of the

reference level approach for the generation of efficient energy supply strategies, in: *Multiobjective and Stochastic Optimization*, Grauer, M., Lewandowski, A. and Wierzbicki, A.P., Eds., IIASA Collaborative Proceedings Series CP-82-S12, Laxenburg, 1982.

17. Henig, M.I., Proper efficiency with respect to cones, *Journal of Optimization Theory and Applications*, 36, 387, 1982.

18. Holmes, R.B., *Geometric Functional Analysis and its Applications*, Springer, New York, 1975.

19. Huang, S.C., Note on the mean-square strategy of vector values objective function, *Journal of Optimization Theory and Applications*, 9, 364, 1972.

20. Hurwicz, L., Programming in linear spaces, in: *Studies in Linear and Non-Linear Programming*, Arrow, K.J., Hurwicz, L. and Uzawa, H., Eds., Stanford University Press, Stanford, 1958.

21. Jahn, J., Duality in vector optimization, *Mathematical Programming*, 25, 343, 1983.

22. Jahn, J., Zur vektoriellen linearen Tschebyscheff-Approximation, *Mathematische Operationsforschung und Statistik, Series Optimization*, 14, 577, 1983.

23. Jahn, J., Scalarization in vector optimization, *Mathematical Programming*, 29, 203, 1984.

24. Jahn, J., A characterization of properly minimal elements of a set, *SIAM Journal on Control and Optimization* (to appear).

25. Kantorovitch, L., The method of successive approximations for functional equations, *Acta Mathematica*, 71, 63, 1939.

26. Kitagawa, H., Watanabe, N., Nishimura, Y. and Matsubara, M., Some pathological configurations of noninferior set appearing in multicriteria optimization problems of chemical processes, *Journal of Optimization Theory and Applications*, 38, 541, 1982.

27. Koller, R. and Farwick, H., Optimierung von Konstruktionen mit Hilfe elektronischer Datenverarbeitungsanlagen, *Industrie-Anzeiger*, 96, 158, 1974.

28. Krabs, W., *Optimization and Approximation*, John Wiley & Sons, Chichester, 1979.

29. Krein, M.G. and Rutman, M.A., Linear operators leaving invariant a cone in a Banach space, in: *Functional Analysis and Measure Theory*, AMS Translation Series 1, Volume 10, Providence, 1962.

30. Lin, J.G., Maximal vectors and multi-objective optimization, *Journal of Optimization Theory and Applications*, 18, 41, 1976.

31. Pascoletti, A. and Serafini, P., Scalarizing vector optimization problems, *Journal of Optimization Theory and Applications*, 42, 499, 1984.

32. Reemtsen, R., On level sets and an approximation problem for the numerical solution of a free boundary problem, *Computing*, 27, 27, 1981.

33. Rolewicz, S., On a norm scalarization in infinite dimensional Banach

spaces, *Control and Cybernetics*, 4, 85, 1975.

34. Salukvadze, M.E., Optimization of vector functionals (in Russian), *Automatika i Telemekhanika*, 8, 5, 1971.

35. Stadler, W., Applications of multicriteria optimization in engineering and in the sciences (a survey), forthcoming.

36. Steuer, R.E. and Choo, E.-U., An interactive weighted Tchebycheff procedure for multiple objective programming, *Mathematical Programming*, 26, 326, 1983.

37. Vogel, W., Ein Maximum-Prinzip für Vektoroptimierungs-Aufgaben, *Operations Research Verfahren*, XIX, 161, 1974.

38. Vogel, W., *Vektoroptimierung in Produkträumen*, Verlag Anton Hain, Meisenheim am Glan, 1977.

39. Vogel, W., Halbnormen und Vektoroptimierung, in: *Quantitative Wirtschaftsforschung - Wilhelm Krelle zum 60. Geburtstag*, Tübingen, 1977.

40. Wierzbicki, A.P., Penalty methods in solving optimization problems with vector performance criteria, Technical Report of the Institute of Automatic Control, TU of Warsaw, 1974.

41. Wierzbicki, A.P., Basic properties of scalarizing functionals for multiobjective optimization, *Mathematische Operationsforschung und Statistik, Series Optimization*, 8, 55, 1977.

42. Wierzbicki, A.P., The use of reference objectives in multi-objective optimization, in: *Multiple Criteria Decision Making - Theory and*

Application, Fandel, G. and Gal, T., Eds., Springer, Berlin, 1980.

43. Wierzbicki, A.P., A mathematical basis for satisficing decision making, in: *Organizations: Multiple Agents with Multiple Criteria*, Morse, J.N., Ed., Springer, Berlin, 1981.

44. Yu, P.L., A class of solutions for group decision problems, *Management Science*, 19, 936, 1973.

45. Yu, P.L. and Leitmann, G., Compromise solutions, domination structures, and Salukvadze's solution, *Journal of Optimization Theory and Applications*, 13, 362, 1974.

A UNIFIED APPROACH FOR SCALAR AND VECTOR OPTIMIZATION

Paolo Serafini
Dipartimento di Matematica ed Informatica
Università di Udine

1. INTRODUCTION

This paper deals with some concepts related to the theory of convex programming. A theoretical framework is developed where both scalar and vector optimization can be accomodated. So far vector optimization the adopted point of view is much in the spirit of scalarization ; in this sense it is closely related to the papers by Pascoletti and Serafini[1] and Jahn[2,3]. Moreover it develops in a more general way ideas first appeared in Serafini[4].

The main attempt of this paper is to derive some basic facts from a few essential ingredients, i.e. a set and a sublinear functional defined in the space of the objectives and/or of the constraints. Starting from this it is possible to derive a cone ordering and to verify a saddle relationship which leads directly to a dual interpretation of the original problem. These general facts are then specialized. It is seen that one particular type of functional allows for a compact characterization of both vector optimization problems and scalar constrained optimization problems.

2. MATHEMATICAL PRELIMINARIES

We shall be concerned with the two following objects :

a) a set $K \subset R^n$

b) a sublinear closed functional $p : R^n \to R \cup \{\infty\}$

We recall that a sublinear functional is defined by the following properties: $p(\alpha x)=\alpha p(x)$ for any $\alpha>0$; $p(x_1+x_2)\leq p(x_1)+p(x_2)$; $p(0)=0$; and that a functional p is closed iff the sets $\{x:p(x)\leq\alpha\}$ are closed for any α.

Our aim will be the minimization of p over K. We shall need the following:

Assumption 1

There exists an open convex cone C such that :

$K \subset C$ and $p(x) > 0$ for any x in C, except possibly at the origin. ∎

The cone C can be viewed to as the "environment" of p and K. We shall also assume the following properness condition:

Assumption 2

There exists x in K such that $p(x) < \infty$. ∎

Then the closedness of p is equivalent to lower semicontinuity.

We may associate to p its dual functional p* :

$$p^*(\pi) = \sup_{\substack{p(x)\le 1 \\ x\in C}} \pi x$$

We note that, by dropping the condition $x\in C$, p* would be the polar functional of p (see Rockafellar[5] p. 128).

Proposition 1

$p^* : R^n \to R \cup \{\infty\}$ is a nonnegative closed sublinear functional. (the proof is straightforward) ∎

Let us remark that, as it follows directly from the definition, :

$$\pi x \le p(x)\, p^*(\pi) \quad \text{for any } x \text{ in } C \text{ and any } \pi .$$

This inequality resembles the familiar Cauchy-Schwarz inequality, and therefore we shall refer to it as the extended Cauchy-Schwarz inequality. It is known that the Hahn-Banach theorem allows for the existence of a π, for any fixed x, such that the Cauchy-Schwarz inequality is satisfied as an equality. Hence we are asking whether something similar can be stated for the extended Cauchy-Schwarz inequality. With this regard we have the following:

Proposition 2

For any x in C, $x\neq 0$, there exists a $\pi\neq 0$ such that :

a) if $p(x) < \infty$ then $\pi x = p(x)\, p^*(\pi)$

b) if $p(x) = \infty$ then $p^*(\pi) = 0$ and $\pi x \ge \pi' x$ for any π' such that $p^*(\pi')=0$ and $\|\pi'\| = \|\pi\|$

Proof

a) Let $A=\{y\in C: p(y)<p(x)\}$. A is nonempty and convex. Hence there exists $\pi\neq 0$ such that $\pi x \ge \pi y$ for any y in A, or alternatively, $\pi x \ge p(z)\ \pi z$ for any z such that $p(z)<1$. By taking the sup for all z in C and $p(z)>1$ we get, since p is closed, $\pi x \ge p(x)\, p^*(\pi)$. This combined with the extended Cauchy-

Schwarz inequality proves the thesis a).

b) Let $A = \{y : p(y) < \infty\}$. A is nonempty and convex. Let S_ε be the open ball of radius ε centered at x (define $S_0 = \{x\}$). Let $\alpha = \sup\{\varepsilon : S_\varepsilon \cap A = \emptyset\}$. Then there exists a π with norm unity such that $\pi z \geq \pi y$ for any z in S_α and any y in A. Also $\pi y \leq 0$ otherwise πy would be unbounded above and therefore $p^*(\pi)=0$. Similarly $\pi z \geq 0$ (since $p(z)=\infty$ in S_α).

Now suppose there exists $\hat{\pi}$ with norm unity such that $p^*(\hat{\pi})=0$ and $\hat{\pi}x > \pi x$, say $\hat{\pi}x = \pi x + \eta + \mu$ with $\eta>0, \mu>0$ arbitrary. Note that $\pi x \geq \alpha$. Hence $\hat{\pi}x \geq \alpha+\eta+\mu \geq (\alpha+\eta+\mu)\, \hat{\pi}u$ for all u with norm not greater than unity, i.e. $\hat{\pi}(x-(\alpha+\eta)u) \geq \mu\, \hat{\pi}u$. Since there is always a u such that $\hat{\pi}u > 0$, we have $\hat{\pi}(x-(\alpha+\eta)u) > 0$ for that u. Now $x-(\alpha+\eta)u$ is in $S_{\alpha+\eta}$ and the previous strict inequality together with $p^*(\hat{\pi})=0$, i.e. $\hat{\pi}y \leq 0$ for any y in A, imply $x-(\alpha+\eta)u$ does not belong to A, which contradicts the fact that $S_{\alpha+\eta} \cap A \neq \emptyset$ for any arbitrary positive η . ∎

Proposition 2 associates to any x in C a set Π_x of π's. Actually Π_x is a convex cone as it is not difficult to see. Let Λ^* be the smallest closed convex cone such that

$$\Lambda^* \supset \bigcup_{x \in C} \Pi_x$$

Therefore for any x in C the associated π is an element of Λ^*.

We are now in the position of defining the dual functional p^{**} of p^*:

$$p^{**}(x) = \sup_{\substack{p^*(\pi) \leq 1 \\ \pi \in \Lambda^*}} \pi x$$

As maybe expexted p^{**} is closely related to p. In fact p and p^{**} agree on C as the following proposition shows :

<u>Proposition 3</u>

$p^{**}(x) = p(x)$ for any x in C

Proof

By Proposition 2, if x is in C

$$p^{**}(x) = \sup_{\substack{p^*(\pi)\leq 1 \\ \pi \in \Lambda^*}} \pi x = \sup_{p^*(\pi)\leq 1} \pi x \quad .$$

Hence p^{**} is also the polar functional of p^*. If we define

$$\hat{p}(x) = \begin{cases} p(x) & \text{for } x \text{ in } \bar{C} \text{ (i.e. the closure of C)} \\ \infty & \text{otherwise} \end{cases}$$

we have that p^* is the polar of $\hat{p}$ and therefore p^{**} is the polar of the polar of $\hat{p}$. Since $\hat{p}$ is closed we have $p^{**} = \hat{p}$ (see Rockafellar[5] p. 128). $\blacksquare$

In order to simplify the notation let us now define:

$$\Pi = \{ \ \pi \in \Lambda^* : p^*(\pi) \leq 1 \ \}$$

3. MONOTONICITY PROPERTIES

Let $f: X \to R$ and $A \subset X$. Following Jahn[2,3] we define :

a) f is monotonic with respect to A if $a \in A$ implies $f(x+a) \geq f(x)$;

b) f is strictly monotonic with respect to A if it is monotonic and $a \in \text{int } A$ implies $f(x+a) > f(x)$.

Let us now consider the following cone:

$$\Lambda = \{ \ \lambda : \pi\lambda \geq 0 \ \text{ for any } \pi \text{ in } \Lambda^* \ \}$$

The cone Λ and the functional p are closely related by the following proposition:

Proposition 4

p is strictly monotonic with respect to Λ over C.

Proof

If $\lambda \in \Lambda$ $\quad p(x+\lambda) \geq p(x+\lambda)p^*(\pi) \geq \pi(x+\lambda) \geq \pi x$ for any $\pi \in \Pi$. Hence

$$p(x+\lambda) \geq \sup_{\pi\in\Pi} \pi x = p^{**}(x) = p(x) .$$

This establishes monotonicity. In order to show strict monotonicity let $\lambda \in \text{int}\,\Lambda$. We shall consider two cases according whether the sup in the definition of p^{**} is achieved or not:

a) let $p^{**}(x) = \hat{\pi}x$. Then $p(x)=p^{**}(x)=\hat{\pi}x<\hat{\pi}(x+\lambda)\leq\sup_{\pi\in\Pi}\pi(x+\lambda)=p^{**}(x+\lambda)=p(x+\lambda)$

b) since $\lambda\in\text{int}\,\Lambda$ there exists $\varepsilon>0$ such that $\pi\lambda \geq \varepsilon\|\pi\|$. Note that $0\in\Pi$ and therefore as πx tends to $p^{**}(x)$ we are assured that $\|\pi\|$ is bounded below by a positive constant, say α, otherwise the sup would be achieved.

Hence $p(x)=p^{**}(x)= \sup_{\pi\in\Pi} \pi x = \sup_{\substack{\pi\in\Pi\\ \|\pi\|\geq\alpha}} \pi x < \sup_{\substack{\pi\in\Pi\\ \|\pi\|\geq\alpha}}(\pi x+\varepsilon\alpha) \leq \sup_{\substack{\pi\in\Pi\\ \|\pi\|\geq\alpha}} (\pi x+\pi\lambda) \leq$

$$\sup_{\pi\in\Pi} (\pi x+\pi\lambda) = p^{**}(x+\lambda) = p(x+\lambda)$$ ∎

4. SADDLE POINTS

As previously remarked we are mainly concerned with the problem

$$v = \inf_{x\in K} p(x)$$

By Proposition 3 this problem can be rewritten as

$$v = \inf_{x\in K} p(x) = \inf_{x\in K} p^{**}(x) = \inf_{x\in K} \sup_{\pi\in\Pi} \pi x$$

Quite naturally we may define the dual problem as

$$d = \sup_{\pi\in\Pi} \inf_{x\in K} \pi x$$

and, by defining the support functional of K as

$$h(\pi) = \inf_{x\in K} \pi x$$

the dual problem can be rewritten as

$$d = \sup_{\pi \in \Pi} h(\pi)$$

It is well known that $d \leq v$. We are now interested in finding conditions which guarantee the existence of a saddle value, i.e. $d = v$, and the existence of a saddle point, i.e. $d = v = \pi x$ with $x \in K$ $\pi \in \Pi$ .Clearly these conditions will concern Π and K only. By construction Π is closed and convex.

Let us first note that convexity of K is not enough to guarantee existence of a saddle value. In fact we have the following proposition which is a well known result (see Rockafellar[5] p. 393 and p. 397) :

<u>Proposition 5</u>

Let K be closed and convex. If either K or Π is bounded a saddle value exists (finite by Assumptions 1 and 2). If both are bounded then there exists also a saddle point. ∎

However the hypotheses required in Proposition 5 are too strong. In fact note that, since $\inf\{p(x):x\in K\} = \inf\{p(x):x\in K+\Lambda\}$ by Proposition 4 , one would expect existence of a saddle value also in the presence of the unbounded set $K+\Lambda$. Besides Π is unbounded iff there exists $\pi\neq 0$ such that $p^*(\pi)=0$, and the latter condition is implied by the presence of some $x\in C$ such that $p(x)=\infty$. As we shall see, this possibility is very important.

Therefore we are looking for weaker hypotheses guaranteeing existence of a saddle value and a saddle point. Let us define:

$$\Lambda_0^* = \{ \pi \in \Lambda^* : p^*(\pi)=0 \}$$

We recall that :

i) the recession cone 0^+A of a convex set A is the largest cone such that $A + 0^+A = A$ (hence $0^+A = \{0\}$ for a bounded set A). Note that $0^+\Pi = \Lambda_0^*$.

ii) ri A is the relative interior of A, i.e. the set of interior points in the relative topology induced by the smallest affine subspace containing A.

We may now state the following assumptions which we call qualifications, as it will become clear later:

Qualifications

Q1 p^* is locally bounded

Q2 for any $x \in 0^+K$ $x \neq 0$ there exists $\pi \in \text{ri}\,\Pi$ such that $\pi x > 0$;

Q3 there exists a neighbourhood N such that $(K+z) \cap \{x : p(x) < \infty\} \neq \emptyset$ $\forall z \in N$;

Q4 for any $\pi \neq 0$ $\pi \in \Lambda_o^*$ there exists $x \in \text{ri}\,K$ such that $\pi x < 0$. ∎

Note that Q1 and Q2 are implied by boundedness of K,and Q3, Q4 are implied by boundedness of Π.

By using qualifications we may state a weaker version of Proposition 5:

Proposition 6

Let K be closed and convex. If any one of the qualifications Q1-Q4 holds, then there exists a saddle value (finite by Assumptions 1 and 2). If either Q1 or Q2 and either Q3 or Q4 hold, then there exists a saddle point.

(the proof can be worked out by using the results in Rockafellar[5], Ch.37) ∎

It is useful to briefly comment the qualifications:

i) note that an equivalent way to state Q2 is the following

$$0^+K \cap (-\Lambda) = \{0\}$$

This amounts to requiring well behaviour of p and K with respect to minimization, in the sense that unboundedness is only required along directions of decrement for p. Note that Assumption 1 and monotonicity imply $-\Lambda \cap C = \emptyset$. Since $0^+K \subset \bar{C}$, it would suffice $-\Lambda \cap \bar{C} = \{0\}$ for Q2 to hold.

It is important to note that (for a convex K) $0^+K \cap (-\Lambda) = \{0\}$ guarantees $0^+(K+\Lambda) = 0^+K + 0^+\Lambda = 0^+K + \Lambda$ and consequently $0^+(K+\Lambda) \cap (-\Lambda) = \{0\}$ (Rockafellar[5] p. 75). In other words if Q2 holds for K then it does for $K+\Lambda$ as well.

ii) note that Q1 implies Q2 since $-\Lambda \cap \bar{C} \neq 0$ implies p^* locally unbounded,

iii) Q3 is a regularity assumption for a set intersection, in other words the two sets have to be in a generic position in order to have a nonempty intersection for any sufficiently small displacement;

iv) note that $p^*(\pi)=0$ implies $\pi x \leq 0$ for any x such that $p(x)<\infty$. Hence Q4 requires at least one point of K to be in tne interior of the set $\{x:p(x)<\infty\}$;

v) it is not difficult to see that Q3 and Q4 are equivalent.

We may therefore conclude that Q1 and Q2 are reasonable assumptions usually met in most applications, whilst Q3 and Q4 have to be carefully considered as we shall see. However it is important to stress that Q1-Q4 are sufficient conditions, that is saddle points could exist without Q1-Q4 being satisfied. As an example consider the case of polyhedral K and Λ for which saddle points are guaranteed without assuming qualifications.

5. SPECIALIZING THE FUNCTIONAL p

Let us now consider the previous results for some particular p.

a) $p(x) = \|x\|$

$C = R^n$; K is any closed convex set not containing the origin.

Then obviously $p^*(\pi) = \|\pi\|$, $\Lambda^* = R^n$, $\Lambda = \{0\}$. Hence monotonicity is trivially satisfied and Q1-Q4 clearly hold. Therefore

$$\min_{x \in K} \|x\| = \max_{\|\pi\| \leq 1} h(\pi)$$

which is a well known result (Luenberger[6] p. 136);

b) $p(x) = p\, x$

$C = \{x:px>0\}$ (open half space) ; K is any closed convex set in C .

Then $p^*(\pi) = \begin{cases} \alpha & \text{for } \pi = \alpha p \quad \alpha \geq 0 \\ 0 & \text{for } \pi = \alpha p \quad \alpha \leq 0 \\ \infty & \text{otherwise} \end{cases}$

Hence $\pi x = p\, x\, p^*(\pi)$ can be satisfied only if $\pi = \alpha p$ $\alpha > 0$. Therefore $\Lambda^* = \{\pi : \pi = \alpha p\ \alpha \geq 0\}$, $\Lambda = \bar{C}$. Monotonicity is easily verified and the saddle point relationship becomes trivial in this case.

Examples a) and b) were rather simple. In fact the cone C was in both cases the largest allowable cone for the given functionals, whereas it is only necessary that $C \supset K$. With this idea in mind we elaborate the previous examples.

c) $p(x) = \|x\|$ $C \neq R^n$

Hence we have $p^*(\pi) \leq \|\pi\|$ in general. Let $x \to \pi_x$ the correspondence which associates to x a π_x such that $\pi_x x = \|x\|\,\|\pi_x\|$. Hence :

$p^*(\pi_x) p(x) \leq \|\pi_x\|\ p(x) = \|\pi_x\|\ \|x\| = \pi_x x \leq p^*(\pi_x)\ p(x)$ and therefore

$\pi_x x = p^*(\pi_x)\ p(x)$ and $p^*(\pi_x) = \|\pi_x\|$. It can be also proven that, conversely $\pi x = p(x)\ p^*(\pi)$ implies $\pi = \pi_x$.

Clearly restricting C has the effect of restricting Λ^* and consequently widening Λ. Consider for instance the case of a Chebyshev norm with C the positive orthant. Then $\Lambda^* = R^n_+ = \Lambda$ and $p^*(\pi) = \|\pi\|_1$.

As an immediate result we obtain the known fact that the Chebyshev norm is strictly monotonic with respect to the Pareto ordering (note that convexity is not called for).

The saddle relationship for closed convex K specializes to

$$\min_{x \in K} \|x\|_\infty = \max_{\substack{\Sigma \pi_i \leq 1 \\ \pi_i \geq 0}} h(\pi)$$

This last example leads directly to a more general setting :

d) $p(x) = \inf\{ p : x - p\, q \in -\Gamma \}$ ($=\infty$ if no such p exists)

with Γ closed convex cone, $q \in \Gamma \cap C$.

It is easy to see that $q \in \operatorname{int} \Gamma$ implies $p(x) < \infty$ for any x .

Let us note that this functional corresponds to the scalarization proposed in Pascoletti and Serafini[1] (with different notation) and that it is very close to the Minkowski functional introduced in Jahn[2,3], the only difference being that for $x \in -\Gamma$ the Minkowski functional vanishes whereas $p(x) \leq 0$. Moreover if Γ is the nonnegative orthant $p(x)$ agrees with the weighted Chebyshev norm for $x \in \Gamma$.

In order to compute p^* it is convenient to note that to any x there is associated a $\gamma(x) \in \Gamma$ (in fact $\gamma(x) \in \partial\Gamma$ the boundary of Γ) such that $\gamma(x) = p(x)q - x$. Hence for $\pi \in \Gamma^*$ one has:

$$p^*(\pi) = \sup_{\substack{p(x) \leq 1 \\ x \in C}} \pi x = \sup_{\substack{p(x) \leq 1 \\ x \in C}} (\pi q\, p(x) - \pi\gamma(x)) \leq \sup_{\substack{p(x) \leq 1 \\ x \in C}} \pi q\, p(x) = \pi q$$

Moreover $p(q) = 1$ hence $\pi q \leq p^*(\pi)$. Therefore $p^*(\pi) = \pi q$ if $\pi \in \Gamma^*$. It is also possible to show that $p^*(\pi) > \pi q$ if $\pi \notin \Gamma^*$.

Now we want to show that $\Lambda^* = \Gamma^*$ (and therefore $\Lambda = \Gamma$) .

Let $\pi \in \Gamma^*$. Take $x = \alpha q$. Then $p(x) = \alpha$. Hence $\pi x = \alpha \pi q = \alpha p^*(\pi) = p(x) p^*(\pi)$ whence $\pi \in \Lambda^*$ i.e. $\Gamma^* \subset \Lambda^*$. In order to show the reverse inclusion we have to consider two cases:

a) $p(x) < \infty$. Let $\pi \in \Lambda^*$. Hence $\pi x = p(x) p^*(\pi)$. Now for any $\gamma \in \Gamma$ there is $\alpha > 0$ such that $x - \alpha\gamma \in C$ and $p(x) \geq p(x - \alpha\gamma)$. Therefore $\pi(x - \alpha\gamma) \leq p(x) p^*(\pi)$ i.e. $\pi\gamma \geq 0$ for any $\gamma \in \Gamma$ which implies $\pi \in \Gamma^*$,

b) $p(x) = \infty$. Then $p^*(\pi) = 0$ implies $\pi \in \Gamma^*$.

Therefore from a) and b) $\Lambda^* \subset \Gamma^*$

Hence p is strictly monotonic with respect to Γ , as expected, and the dual problem becomes:

$$d = \sup \{ h(\pi) : \pi q \leq 1 \ , \ \pi \in \Gamma^* \}$$

6. APPLICATIONS TO MATHEMATICAL PROGRAMMING

In this section we shall devote our attention to the functional p defined in the last example of the previous section. Its particular form will allow a nice application to usual mathematical programming problems:

a) let $f : \Omega \to R^n$. Let R^n be partially ordered by a closed convex cone Λ . We want to find elements $\omega \in \Omega$ such that $f(\omega)$ is minimal.
Let $K = f(\Omega)+\Lambda \neq R^n$. Suppose without loss of generality $0 \notin K$.
Let $P(\omega)=p(f(\omega))$ (with p as in d) of the previous section with $\Gamma=\Lambda$).
Hence in general :

$$\inf_{\omega \in \Omega} P(\omega) \geq \sup_{\substack{\pi q \leq 1 \\ \pi \in \Lambda^*}} \inf_{\omega \in \Omega} \pi f(\omega)$$

We shall denote by Co A the convex hull of a set A.

If we assume $0^+ \text{Co}\, f(\Omega) \cap (-\Lambda) = \{0\}$ then Q1 and Q2 are satisfied (a very natural assumption). Moreover by assuming $q \in \text{int}\Lambda$, Q3 and Q4 are satisfied as well. Therefore we have:

<u>Proposition 7</u>

If f is convex, Ω is convex, $0^+ \text{Co}\, f(\Omega) \cap (-\Lambda) = \{0\}$, K is closed and $q \in \text{int}\, \Lambda$, then

$$P(\omega_o) = \min_{\omega \in \Omega} P(\omega) = \max_{\substack{\pi q \leq 1 \\ \pi \in \Lambda^*}} \inf_{\omega \in \Omega} \pi f(\omega) = \pi_o f(\omega_o) = d(q)$$

■

Since by definition $f(\omega_o) - P(\omega_o)\, q \in -\Lambda$ and $f(\omega_o) - p\, q \notin -\Lambda$ for any $p < P(\omega_o)$, one gets $P(\omega_o)\, q = d(q)\, q \in \partial K$ and obviously

$$\bigcup_{q \in \text{int}\Lambda} \{ d(q)\, q\} = \partial K \cap \text{int}\, \Lambda$$

In other words, by varying q it is possible to explore the set of minimal points by using the dual problem.

In case of no convexity of f or Ω then

$$d = \inf_{x \in \mathrm{Co}\,K} p(x) \leq \inf_{x \in K} p(x) = \inf_{\omega \in \Omega} P(\omega)$$

and $\quad d\,q \in \partial \mathrm{Co}\,K$. If we define:

<u>Definition</u>

Let R^n be partially ordered by a cone Λ. Let A be a set in R^n. Then b is a lower bound for A if $b-a \notin \Lambda\setminus\{0\}$ for any $a \in A$. A set B is a lower bound for A if all its elements are lower bounds for A, that is if $B \cap (A+\Lambda\setminus\{0\}) = \emptyset$.

Moreover b is a weak lower bound (w.l.b.) if $b-a \notin \mathrm{int}\Lambda$ for any $a \in A$, and B is a w.l.b. if $B \cap (A+\mathrm{int}\Lambda) = \emptyset$.

Then clearly $d(q)\,q$ is a w.l.b. for the set of minimal points and there is always a cone $Q \subset \Lambda$ such that $d(q)\,q$ is a l.b. for all $q \in Q$.

b) Let $\quad f : \Omega \to R$

$\qquad g : \Omega \to R^m$

Let R^{m+1} be naturally ordered ; let $\quad K = \begin{pmatrix} f \\ g \end{pmatrix}(\Omega) + R_+^{m+1}$

Let $\Gamma = R_+^{m+1}$ and $q=(1,0,\dots,0)$ in the definition of p.

Then the problem $\inf_{\omega \in \Omega} P(\omega)$ is equivalent to the problem $\quad v = \inf_{\substack{g(\omega)\leq 0 \\ \omega \in \Omega}} f(\omega)$

Notice that with this particular definition of q, replacing $f(\omega)$ with $f(\omega)+c$ has the effect of replacing $P(\omega)$ with $P(\omega)+c$.

Let us assume that $-\infty<v<\infty$ and that $\mathrm{Co}\,K \neq R^{m+1}$. These are very natural assumptions which guarantee the existence of a constant c such that both $v+c>0$ and the support function of $K+\binom{c}{0}$ is positive for some arguments.

Let us now consider the dual problem with respect to the new function $f(\omega)+c$. Hence

$$d' = \sup_{\substack{\pi q \le 1 \\ \pi \ge 0}} \inf_{\omega \in \Omega} (\pi_o(f(\omega)+c)+\pi_1 g(\omega)) = \sup_{\substack{0 \le \pi_o \le 1 \\ \pi_1 \ge 0}} (\pi_o c + h(\pi))$$

By the previous assumptions there exist π such that $\pi_o c+h(\pi) >0$. This implies $\pi_o \neq 0$, otherwise $d'=\infty$ whilst $d' \le v+c$. Moreover by positive homogeneity of $\pi_o c+h(\pi)$ one gets $\pi_o=1$. Therefore one derives the well known duality formulation through the Lagrangian :

$$d' = c + \sup_{\pi_1 \ge 0} \inf_{\omega \in \Omega} (f(\omega)+\pi_1 g(\omega)) = c + d \le c + v$$

Since Π is unbounded in this case, the existence of saddle points is guaranteed by convexity and Q3, which is nothing but the Slater's constraint qualification.

c) Let $f: \Omega \to R^n$

$g: \Omega \to R^m$

Let $\Lambda \subset R^n$ be a closed convex cone.

Then the problem of finding points ω such that $f(\omega)$ is minimal with respect to Λ and $g(\omega) \le 0$ can be solved by combining together the previous cases a) and b). Therefore let us define :

$$p(x) = \inf \{ p : x - p \binom{q}{0} \in \binom{-\Lambda}{R^m_-} \} \qquad K = \binom{f}{g}(\Omega) + \binom{\Lambda}{R^m}$$

As before note that $p(x_o) \le p(x)$ for any $x \in K$ implies $x_o \in \partial K$ and $p(x_o)\binom{q}{0} \in \partial K$. The dual problem becomes :

$$d = \sup_{\substack{\pi_o q \le 1 \\ \pi_o \in \Lambda^* \\ \pi_1 \ge 0}} \inf_{\omega \in \Omega} (\pi_o f(\omega) + \pi_1 g(\omega))$$

and clearly, since $d \le p(x_o)$, $d \binom{q}{0}$ is a w.l.b. for ∂K, in fact $d \binom{q}{0} \in \partial \mathrm{Co}\, K$.

Let $X = \{ x \in R^n : x - f(\omega) \in \Lambda \;,\; g(\omega) \leq 0 \;,\; \omega \in \Omega \}$. Then ∂Co X is a w.l.b. for the set of minimal points $f(\omega)$.
Let $Z = \{ x = (x_o, x_1) \in \text{Co } K \subset R^{n+m} : x_1 \leq 0 \}$. Let Y denote the natural projection of Z into R^n. Then $d \binom{q}{0} \in \partial\text{Co } K$ implies $d\, q \in \partial Y$. We want to show that ∂Y is a w.l.b. for ∂Co X. For notice that $Y+\Lambda=Y$ and therefore $y \in \partial Y$ implies $y-\lambda \notin Y$ for any $\lambda \in \text{int}\,\Lambda$. Since $\text{Co } X \subset Y$, as it is easy to see, also $y-\lambda \notin \text{Co } X$ for any $\lambda \in \text{int}\,\Lambda$, which proves the claim.
The set $Y \setminus X$ represents the duality gap.

7. CONCLUSIONS

It has been shown that several different results can be embedded in a unified theory which focuses especially on duality. All results have been proven by exploiting finite dimensionality properties. We conjecture however that the proofs can be modified to account for more general infinite dimensional spaces.

The properties of the sublinear functional have been decisive in deriving almost all results. In this theoretical framework there is no conceptual difference between objective functions and constraint functions. For a particular type of functional the technical difference consists in the value of the corresponding component of the parameter q. In other words the modelling of objective functions as constraints is already embedded in the problem.

The saddle relationship may turn out to be useful even for non convex problems. In fact the dual problem provides lower bounds and these can be successfully used for methods of a branch-and-bound type.

REFERENCES

1. Pascoletti, A. and Serafini, P., Scalarizing vector optimization problems, *Journal of Optimization Theory and Applications*, 42, 499, 1984.
2. Jahn, J., Scalarization in vector optimization, *Mathematical Programming*, 29, 203, 1984.
3. Jahn, J., Scalarization in multi objective optimization, *this volume.*
4. Serafini, P., Dual relaxation and branch-and-bound techniques for multi objective optimization, in *Interactive decision analysis*, Grauer, M., Wierzbicki, A., Eds., Springer, Berlin, 1984.
5. Rockafellar, R.T., *Convex analysis,* Princeton University Press, 1972.
6. Luenberger, D.G., *Optimization by vector space methods*, Wiley, New York, 1969.

LAGRANGE DUALITY AND ITS GEOMETRIC INTERPRETATION

Hirotaka Nakayama
Department of Applied Mathematics
Konan University

Abstract

In recent years, there have been several reports on duality in vector optimization. However, there seem to be no unified approach to dualization. In the author's previous paper, a geometric consideration was given to duality in nonlinear vector optimization. In this paper, Lagrange duality of vector optimization will be overviewed along with a geometric interpretation. Moreover, Isermann's duality in linear cases will be derived on the basis of the stated geometric consideration.

1. Introduction

Let X be a set of alternatives in an n-dimensional Euclidean space R^n, and let $f=(f_1,\ldots,f_p)$ be a vector-valued criterion function from R^n into R^p. For given two vectors y^1 and y^2 and a pointed cone K, the following notations for cone-order will be used:

$$y^1 \leqq_K y^2 \iff y^2 - y^1 \in K$$

$$y^1 \leq_K y^2 \iff y^2 - y^1 \in K\backslash\{0\}$$

$$y^1 <_K y^2 \iff y^2 - y^1 \in \text{int } K$$

Furthermore, the K-minimal and the K-maximal solution set of Y are defined, respectively, by

$$\text{Min}_K\, Y := \{\bar{y} \in Y \mid \text{ there is no } y \in Y \text{ such that } y \leq_K \bar{y}\}$$

$$\text{Max}_K\, Y := \{\bar{y} \in Y \mid \text{ there is no } y \in Y \text{ such that } y \geq_K \bar{y}\}.$$

Throughout this paper, for any cone K in R^p we denote the positive dual cone of K by K^o, that is,

$$K^o := \{p \in R^p \mid \langle p,q \rangle \geq 0 \quad \text{for any } q \in K\}$$

where $\langle p,q \rangle$ denotes the usual inner products of p and q, i.e., $p^T q$. For a K-convex set Y, a K-minimal solution $\bar{y}$ is said to be proper, if there exists $\mu \in \text{int } K^o$ such that

$$\langle \mu, y \rangle \geq \langle \mu, \bar{y} \rangle \quad \text{for all } y \in Y.$$

Then, a general type of nonlinear vector optimization may be formulated as follows:

(P) D-minimize $f(x)$ subject to $x \in X$,

where

$$X := \{x \in X' \mid g(x) \leqq_Q 0, \quad X' \subset R^n\}.$$

In the following, we shall review several results regarding duality of vector optimization and show a geometric approach to Isermann duality in linear cases.

2. Linear Cases

Let D, Q and M be pointed convex polyhedral cones in R^p, R^m and R^n, respectively. This means, in particular, that $\text{int } D^o \neq \emptyset$. Isermann[4] has given an attractive dualization in linear cases. In the following, we shall consider it in an extended form.

(P_I): D-minimize $\{Cx: \ x \in X\}$

where

$$X := \{x \in M: \ Ax \geq_Q b\}.$$

(D_I): D-maximize $\{Ub: \ U \in \mathcal{U}_0\}$

where

$$\mathcal{U}_0 := \{U \in R^{p\times m} | \text{ there exists } \mu \in \text{int } D^o \text{ such that } U^T\mu \in Q^o \text{ and } A^T U^T \mu \leqq_{M^o} C^T\mu\}.$$

Let $\text{Min}_D\ (P_I)$ and $\text{Max}_D\ (D_I)$ denote the set of D-minimal value in the primal objective space of (P_I) and that of D-maximal value in the dual objective space of (D_I), respectively. The following duality properties are simple extensions of Isermann[4].

Theorem 2.1

(i) $Ub \not\geq_D Cx$ for all $(U,x) \in \mathcal{U}_0 \times X$.

(ii) Suppose that $\bar{U} \in \mathcal{U}_0$ and $\bar{x} \in X$ satisfy

$$\bar{U}b = C\bar{x}. \tag{2.1}$$

Then $\bar{U}$ is a D-maximal solution to the dual problem (D_I) and $\bar{x}$ is a D-minimal solution to the primal problem (P_I).

(iii) $\text{Min}_D\,(P_I) = \text{Max}_D\,(D_I)$.

Proof: (i): Suppose, to the contrary, that there exist some $\bar{x} \in X$ and $\bar{U} \in \mathcal{U}_0$ such that

$$\bar{U}b \geq_D C\bar{x}. \tag{2.2}$$

Note here by definition that there exists $\bar{\mu} \in \text{int } D^o$ such that

$$(\bar{U}A)^T\bar{\mu} \leq_{M^o} C^T\bar{\mu}^T$$
$$\bar{U}^T\bar{\mu} \in Q^o.$$

Therefore, since $\bar{x} \in M$, we have

$$\langle\bar{\mu},\ \bar{U}A\bar{x}\rangle \leq \langle\bar{\mu},\ C\bar{x}\rangle \tag{2.3}$$

Furthermore, from (2.2)

$$\langle\bar{\mu},\ \bar{U}b\rangle > \langle\bar{\mu},\ C\bar{x}\rangle. \tag{2.4}$$

It then follows from (2.3) and (2.4) that

$$\langle\bar{\mu},\ \bar{U}A\bar{x}\rangle < \langle\bar{\mu},\ \bar{U}b\rangle. \tag{2.5}$$

On the other hand, since $\bar{U}^T\mu \in Q^o$ and $A\bar{x} \geq_Q b$, we have

$$\langle\bar{\mu}^T,\ \bar{U}A\bar{x}\rangle \geq \langle\bar{\mu},\ \bar{U}b\rangle,$$

which contradicts (2.5).

(ii): Suppose, to the contrary, that $\bar{U}b \notin \text{Max}_D\,(D_I)$. Then there exists $\hat{U} \in \mathcal{U}_0$ such that

$$\hat{U}b \geq_D \bar{U}b = C\bar{x}$$

which contradicts the result (i). Therefore, $\bar{U}$ is a D-maximal soluiton to the dual problem. In a similar fashion, we can conclude that $\bar{x}$ is a D-minimal solution to the primal problem.

(iii): First we shall prove $\text{Min}_D(P_I) \subset \text{Max}_D(D_I)$. Let $\hat{x}$ be a D-minimal solution to the primal problem (P_I). Then it is readily seen from the well knnown scalarization property that there exists some $\hat{\mu} \in$

int D^o such that

$$\hat{\mu}^T C\hat{x} = \operatorname*{Min}_{x \in X} \hat{\mu}^T Cx.$$

It is sufficient to prove the statement in case where $\hat{x}$ is a basic solution. Transform the original inequality constraints $Ax \geq_Q b$ into

$$Ax - y = b,$$
$$y \geq_Q 0.$$

Let B denote the submatrix of [A, -I] which consists of m columns corresponding to the basic variables. Then from the initial simplex tableu

$$\left(\begin{array}{c|c|c} A & -I & b \\ \hline \hat{\mu}^T C & 0 & 0 \end{array}\right)$$

we obtain the final tableu

$$\left(\begin{array}{c|c|c} B^{-1}A & -B^{-1} & B^{-1}b \\ \hline \hat{\mu}^T(C-C_B B^{-1}A) & \hat{\mu}^T C_B B^{-1} & -\hat{\mu}^T C_B B^{-1}b \end{array}\right)$$

by using the simplex method. According to the well known properties of linear programming, we have

$$(C - C_B B^{-1}A)^T \hat{\mu} \geq_{M^o} 0$$
$$(C_B B^{-1})^T \hat{\mu} >_{Q^o} 0.$$

Setting here $\hat{U} = C_B B^{-1}$, these relations can be rewritten as

$$C^T \hat{\mu} \geq_{M^o} A^T \hat{U}^T \hat{\mu},$$
$$\hat{U}^T \hat{\mu} \in Q^o,$$

which shows that $\hat{U} \in \mathcal{U}_0$.

On the other hand,

$$\hat{U}b = C_B B^{-1} b = C_B x_B = C(x_B, 0) = C\hat{x}.$$

In view of the result (ii), the last relation implies that $\hat{U}$ is a D-maximal solution to the dual problem (D_I). Hence we have $\mathrm{Min}_D\ (P_I)$ $\mathrm{Max}_D\ (D_I)$.

Next, we shall prove the reverse inclusion. Suppose that $\hat{U}$ is a D-maximal solution to the dual problem (D_I). Then it is clear that for every $\mu \in \mathrm{int}\ D^o$ there exists no U with $U^T\mu \in Q^o$ such that

$$A^T U^T \mu \leqq_{M^o} C^T \mu$$
$$\mu^T U b > \mu^T \hat{U} b.$$

Setting here $\lambda = U^T\mu$, it follows that for no $\lambda \in Q^o$ and no $\mu \in \mathrm{int}\ D^o$

$$\begin{aligned} A^T\lambda &\leqq_{M^o} C^T\mu \\ \lambda^T b &> \mu^T \hat{U} b. \end{aligned} \tag{2.6}$$

More strongly, we can conclude that there exist no $\lambda \in Q^o$ and no $\mu \in D^o$ satisfying (2.6). In fact, suppose to the conrary, that there exist some $\lambda' \in Q^o$ and $\mu' \in D^o$ such that

$$A^T\lambda' \leqq_{M^o} C^T\mu'$$
$$\lambda'^T b > \mu'^T \hat{U} b.$$

On the other hand, since $\hat{U}$ is a solution to the dual problem, there exist $\hat{\lambda} = \hat{U}^T\hat{\mu} \in Q^o$ and $\hat{\mu} \in \mathrm{int}\ D^o$ such that

$$A^T\hat{\lambda} \leqq_{M^o} C^T\hat{\mu}$$
$$\hat{\lambda}^T b = \hat{\mu}^T \hat{U}^T b.$$

Therefore, we have

$$A^T(\hat{\lambda} + \lambda') \leqq_{M^o} C^T(\hat{\mu} + \mu')$$
$$(\hat{\lambda} + \lambda')^T b > (\hat{\mu} + \mu')^T \hat{U} b.$$

This implies the existence of solutions $\hat{\mu}+\mu' \in \text{int } D^o$, $\hat{\lambda}+\lambda' \in Q^o$ to (2.6), which is a contradiction.

Rewriting (2.6), we may see that there exist no $\lambda \in Q^0$ and no $\mu \in D^0$ for which

$$\begin{pmatrix} A & -b \\ -C & \hat{\text{Ub}} \end{pmatrix}^T \begin{pmatrix} \lambda \\ \mu \end{pmatrix} \in -M^o \times R^1_+ \setminus \{0\}$$

is satisfied. Thus from Lemma 2.1 below, there exists a $(\hat{x},\alpha) \in$ int $(M^o \times R^1_+)^o$ = int $(M \times R^1_+)$ satisfying

$$\begin{pmatrix} A & -b \\ -C & \hat{\text{Ub}} \end{pmatrix} \begin{pmatrix} \hat{x} \\ \alpha \end{pmatrix} \in Q \times D.$$

Since $\alpha > 0$, we finally have

$$A\hat{x} \geq_Q b \tag{2.7}$$
$$C\hat{x} \leq_D \hat{\text{Ub}}.$$

By using the result (i), the last relation reduces to

$$C\hat{x} = \hat{\text{Ub}}. \tag{2.8}$$

From the result (ii), the relations (2.7) and (2.8) imply that $\hat{x}$ is a D-minimal solution to the primal problem. This completes the proof.

The following lemma, which was used in the proof of Theorem 2.4, is a simple extension of that due to Gale[3].

Lemma 2.1

Either

(I) $A^T\lambda \in -M^o \setminus \{0\}$ for some $\lambda \in Q^o$

or

(II) $Ax \in Q$ for some $x \in \text{int } M$

holds, but never both.

Proof: Suppose, to the contrary, that both (I) and (II) hold simultaneously, then we have $\langle A^T\lambda, x\rangle < 0$ from $A^T\lambda \in -M^o\setminus\{0\}$ and $x \in \text{int } M$, and hence $\langle \lambda, Ax\rangle < 0$. Further, from $b \in Q^o$ and $Ax \in Q$ we have $\langle \lambda, Ax\rangle \geqq 0$, which leads to a contradiction.

Next suppose that (I) does not hold. This implies from Lemma 2.1 that

$$-(AM)^o \cap Q^0 = \{0\}.$$

Hence

$$-AM + Q = R^m,$$

from which there exist $p \in AM$ and $q \in Q$ for some $v \in A(\text{int } M) \subset R^m$ such that

$$v = -p + q.$$

Finally, $q = p + v$ implies

$$q \in Q \cap A(\text{int } M)$$

which also implies (II). This completes the proof.

3. Nonlinear Cases

Let D be a pointed closed convex cone which defines a cone-order $\leqq_D$ as in the previous section. In this section, we shall be concerned with the nonlinear vector optimzation

$$(P): \quad \text{D-minimize} \quad f(x)$$

$$\text{subject to}$$

$$X := \{x \in X' \mid g(x) \leq_Q 0, \quad X' \subset R^n\}.$$

where $f=(f_1,\dots,f_p)$.

For a while in this section, we impose the following assumptions:

(i) X' is a nonempty compact convex set.

(ii) D and Q are pointed closed convex cones with nonempty interior respectively of R^p and R^m.

(iii) f is continuous and D-convex.

(iv) g is continuous and Q-convex.

Under the assumptions, it can be readily shown that for every $z \in R^m$, both sets

$$X(z) := \{x \in X' \mid g(x) \leq_Q z\}$$

and

$$Y(z) := f[X(z)] := \{y \in R^p \mid y=f(x), \quad x \in X', \quad g(x) \leq_Q z\} \qquad (3.1)$$

are compact, $X(z)$ is convex and $Y(z)$ is D-convex. Let us consider the primal problem (P) by embedding it in a family of preturbed problems with $Y(z)$ given by (3.1):

$$(P_z): \quad \text{D-minimize} \quad Y(z).$$

Clearly the primal problem (P) is identical to the problem (P_z) with $z=0$. Now define the set Z as

$$Z := \{z \in R^m \mid X(z) \neq \emptyset\}.$$

It is known that the set Z is convex (see, for example, Luenberger[6]).

Associated with the problem (P), the point-to-set map defined by

$$W(z) := \mathrm{Min}_D\, Y(z)$$

is called a **perturbation (or primal) map**. It is known that (i) for each $z \in Z$, $W(z)$ is a D-convex set in R^p, (ii) the map $W(z)$ is D-monotone, namely,

$$W(z^1) \subset W(z^2) + D$$

for any z^1, $z^2 \in Z$ such that $z^1 \leqq_Q z^2$, (iii) $W(z)$ is a D-convex point-to-set map (Tanino and Sawaragi[11]).

A **vector-valued Lagrangian** function for the problem (P) is defined on X' by

$$L(x,U) = f(x) + Ug(x).$$

Hereafter, we shall denote by $\mathcal{U}$ a family of all $p \times m$ matrices U such that $UQ \subset D$. Such matrices are said to be positive in some literatures (Ritter[9], Corley[3]). Note that for given $\mu \in D^o \setminus \{0\}$ and $\lambda \in Q^o$ there exist $U \in \mathcal{U}$ such that

$$U^T \mu = \lambda.$$

In fact, for some vector e of D with $\langle \mu, e \rangle = 1$,

$$U = (\lambda_1 e,\ \lambda_2 e, \ldots,\ \lambda_m e)$$

is a desired one.

The point-to-set map $\Phi: \mathcal{U} \dashrightarrow \mathcal{P}(R^p)$ defined by

$$\Phi(U) = \mathrm{Min}_D\, \{L(x,U) \mid x \in X'\}$$

is called a **dual map**, where $\mathcal{P}(R^p)$ denotes the power set of R^p. Using this terminology, a dual problem associated primal problem (P) can be defined in parallel with ordinary mathematical programming as follows (Tanino-Sawaragi[11]):

$$(D_T): \qquad \text{D-maximize} \bigcup_{U \in \mathcal{U}} \Phi(U).$$

It is known that (i) for each U, $\Phi(U)$ is a D-convex set in R^p, (ii) $\Phi(U)$ is a D-concave point-to-set map, i.e., for any U^1, $U^2 \in \mathcal{U}$ and any $\alpha \in [0,1]$

$$\Phi(\alpha U^1 + (1-\alpha)U^2) \subset \alpha\Phi(U^1) + (1-\alpha)\Phi(U^2) + D.$$

Proposition 3.1 (Tanino-Sawaragi[11])

If $\hat{x}$ is a proper D-minimal solution to Problem (P), and if the Slater constraint qualification holds, i.e., there exists $\bar{x} \in X'$ such that $g(\bar{x}) <_Q 0$, then there exists a p×m matrix $\hat{U} \in \mathcal{U}$, such that

$$f(\hat{x}) \in \mathrm{Min}_D \{f(x)+\hat{U}g(x) \mid x \in X'\}, \quad \hat{U}g(\hat{x})=0.$$

3.2. Geometric Interpretation of Vector-valued Lagrangian

In Proposition 3.1, we have established that for a properly D-minimal solution $\hat{x}$ to the problem (P) there exists a pxm matrix $\hat{U}$ such that $\hat{U}Q \subset D$ and

$$f(x) + \hat{U}g(x) \not\leq_D f(\hat{x}) \qquad \text{for all } x \in X'$$

under an appropriate regularity condition such as Slater's constraint qualification. The one in ordianry scalar convex optimization corresponding to this theorem asserts the existence of a supporting hyperplane for epi w at $(0,f(\hat{x}))$. On the other hand, in multiobjective optimzation, for the **D-epigraph** of W(z), defined by

$$\text{D-epi } W := \{(z,y) \mid z \in Z, \ y \in R^p, \ y \in W(z)+D\},$$

it geometrically implies the existence of a supporting conical variety (i.e., a translation of cone) supporting D-epi W at $(0,f(\hat{x}))$.

All asumptions on f, g, D and Q of the previous section are

inherited in this section. As is readily shown, D-epi W is a closed convex set in $R^m \times R^p$ under our assumptions. Hereafter we shall assume that the pointed closed convex cone D is polyhedral. Then, as is well known, there exists a matrix M_1 such that

$$D = \{y \in R^p \mid M_1 y \geqq 0\}.$$

Since D is pointed, M_1 has full rank p.

Here define a cone K in $R^m \times R^p$ as

$$K := \{(z,y) \mid M_1 y + M_2 z \leqq 0, \quad z \in R^m, \quad y \in R^p\}. \tag{3.2}$$

For this cone K, the following properties hold: Let the y-intercept of K in $R^m \times R^p$ be denoted by Y_K, namely

$$Y_K := \{y \mid \ (0,y) \in K, \quad 0 \in R^m, \quad y \in R^p\}$$

Then from the definition of M_1, we have $Y_K = -D$. Furthermore, let $\ell(K)$ denote the lineality space of K, i.e., $-K \cap K$. Then, since $\ell(K)$ accords to $\mathrm{Ker}(M_1, M_2)$, it is m-dimensional. For, letting $M_2 = M_1 U$, we have

$$\ell(K) = \{(z,y) \in R^m \times R^p \mid M_1 y + M_2 z = 0\}$$
$$= \{(z,y) \in R^m \times R^p \mid y + Uz = 0\},$$

because the s p matrix M_1 has the maximal rank p. In addition, since the row vectors of M_1 are generators of D^o and $UQ \subset D$, every row vector of M_2 is in Q^o.

Now we can establish a geometric interpretation of Proposition 3.1: We say the cone K supports D-epi W at $(\hat{z},\hat{y})$, if $K \cap ((D\text{-epi } W) - (\hat{z},\hat{y})) \subset \ell(K)$. Clearly, K supports D-epi W at $(\hat{z},\hat{y})$ if and only if

$$M_1(y-\hat{y}) + M_2(z-\hat{z}) \not\leq 0 \quad \text{for all } (z,y) \in D\text{-epi } W.$$

from which

$$y + Uz \not\leq_D f(\hat{x}) + Ug(\hat{x}) \quad \text{for all } (z,y) \in \text{D-epi } W$$

(see, for example, Nakayama[7]). Therefore, we can conclude that Proposition 3.1 asserts the existence of a supporting conical varieties (a translation of K) for D-epi W at $(g(\hat{x}),f(\hat{x}))$. Observe that in ordinary scalar optimization, the suporting conical varieties becomes half spaces, because the lineality is identical to the dimiension of g. This accords with the well known result.

Now we investigate a relation between supporting hyperplanes and supporting conical varieties. Let $H(\lambda,\mu:\gamma)$ be a hyperplane in $R^m \times R^p$ with the normal (λ,μ) such that

$$H(\lambda,\mu:\gamma) := \{(z,y) \mid \langle\mu,y\rangle + \langle\lambda,z\rangle - \gamma = 0, \quad z \in R^m, \quad y \in R^p\}.$$

Define

$$h(\lambda,\mu:\gamma) := \langle\mu,y\rangle + \langle\lambda,z\rangle - \gamma.$$

Then, associated with the hyperlane $H(\lambda,\mu:\gamma)$, several kinds of half spaces are defined as follows:

$$H^+(\lambda,\mu:\gamma) := \{(z,y) \in R^m \times R^p \mid h(\lambda,\mu:\gamma) \geq 0\},$$

$$\overset{\circ}{H}{}^+(\lambda,\mu:\gamma) := \{(z,y) \in R^m \times R^p \mid h(\lambda,\mu:\gamma) > 0\}.$$

H^- and $\overset{\circ}{H}{}^-$ are similarly defined by replacing the inequality > (resp., =) by < (resp., =). In particular, let $H(\lambda,\mu)$ denote the supporting hyperplane for D-epi W with the normal (λ,μ), that is,

$$H(\lambda,\mu) := H(\lambda,\mu:\hat{\gamma}) \quad \text{where} \quad \hat{\gamma} = \sup\{\gamma \mid H^+(\lambda,\mu:\gamma) \supset \text{D-epi } W\}. \tag{3.3}$$

Lemma 3.1 (Nakayama[7])

The linieality space of the cone K given by (3.2) with $M_2=M_1U$ is included in the hyperplane $H(\lambda,\mu:0)$ if and only if the matrix U satisfies $U^T\mu=\lambda$.

Lemma 3.2 (Nakayama[7])

For any supporting hyperplane $H(\lambda,\mu)$ for D-epi W, we have $\lambda \in Q^{o}$ and $\mu \in D^{o}$.

The following lemma clarifies a relation between supporting conical varieties and supporting hyperplanes:

Lemma 3.3 (Nakayama[7])

Let $H(\lambda,\mu)$ be a supporting hyperplane for D-epi W with a supporting point $(\hat{z},\hat{y})$. If $\mu \in \text{int}\, D^{o}$, then any conical variety of the cone K given by (3.2), whose lineality variety passing through $(\hat{z},\hat{y})$ is included in $H(\lambda,\mu)$, supports D-epi W at $(\hat{z},\hat{y})$.

Conversely, if some conical variety of K supports D-epi W at $(\hat{z},\hat{y})$, then there exists a hyperplane $H(\lambda,\mu:\gamma)$ with $\mu \neq 0$ supporting D-epi W at $(\hat{z},\hat{y})$ which contains the lineality variety of the supporting conical variety.

Now we summarize several properties for the support property of conical varieties, saddle points of the vector-valued Lagrangian function, the duality and the unconstrained D-minimization of the vector-valued Lagangian function as the following mutual equivalence theorem:

Proposition 3.2 (Nakayama[7])

The following four conditions are equivalent to one another:

(i) Let M_1 be an s p matrix with full rank whose row vectors generate D^{o} and let M_2 be an $s \times p$ matrix such that $M_1\hat{U} = M_2$ for $\hat{U} \in \mathcal{U}$.

Then $\hat{x}$ solves the primal problem (P) and conical varieties of the cone given by

$$K = \{(z,y) \mid M_1 y + M_2 z \lneq 0, \ y \in R^n, \ z \in R^m\}$$

supports D-epi W at $(0, f(\hat{x}))$.

(ii) $L(\hat{x},\hat{U}) \in \mathrm{Min}_D \{L(x,\hat{U}) \mid x \in X'\} \cap \mathrm{Max}_D \{L(\hat{x},U) \mid U \in \mathcal{U}\}$ and $\hat{U} \in \mathcal{U}$.

(iii) $\hat{x} \in X'$, $\hat{U} \in \mathcal{U}$ and $f(\hat{x}) \in \Phi(\hat{U})$.

(iv) $L(\hat{x},\hat{U}) \in \mathrm{Min}_D \{L(x,\hat{U}) \mid x \in X'\}$, $g(\hat{x}) \lneq_Q 0$ and $\hat{U} g(\hat{x}) = 0$.

Remark 3.1

Note that the condition (iii) means that $f(\hat{x})$ is a D-minimal solution to the primal problem (P) and also a D-maximal solution to the dual problem (D). The solution $(\hat{x},\hat{U})$ satisfying the propety (ii) is called a generalized saddle point.

3.3. Geometric duality

Other approaches to duality in multiobjective optimzation have been given by Jahn[5] and Nakayama[7]. We shall review their results briefly. As in the previous section, the convexity assumption on f and g will be also imposed here, but X' is not necessarily compact.

Define

$$G := \{(z,y) \in R^m \times R^p \mid \ y \geqq_D f(x), \ z \geqq_Q g(x), \ x \in X'\},$$

$$Y_G := \{y \in R^p \mid \ (0,y) \in G, \ 0 \in R^m, \ y \in R^p\}.$$

We restate the primal problem as

(P): D-minimize $\{f(x) \mid \ x \in X\}$,

where

$$X := \{x \in X' \mid \ g(x) \lneq_Q 0, \ X' \subset R^n\}.$$

Associated with this primal problem, the dual problem formulated by Nakayama[7] is as follows:

$$(D_N): \qquad \text{D-maximize} \quad \bigcup_{U \in \mathcal{U}} Y_{S(U)}$$

where

$$Y_{S(U)} := \{y \in R^p \mid \ f(x) + Ug(x) \not\leq_D y, \ \text{ for all } \ x \in X'\}.$$

On the other hand, the one given by Jahn[5] is

$$(D_J): \qquad \text{D-maximize} \quad \bigcup_{\substack{\mu \in \text{int } D^o \\ \lambda \in Q^o}} Y_{H^-(\lambda,\mu)}$$

where

$$Y_{H^-(\lambda,\mu)} := \{y \in R^p \mid \ \langle \mu, f(x)\rangle + \langle \lambda, g(x)\rangle \geqq \langle \mu, y\rangle \quad \text{for all } \ x \in X'\}.$$

Proposition 3.3

(i) For any $y \in \bigcup_{U \in \mathcal{U}} Y_{S(U)}$ and for any $x \in X$,

$$y \not\leq_D f(x).$$

(ii) For any $y \in \bigcup_{\substack{\mu \in \text{int } D^o \\ \lambda \in D^o}} Y_{H^-(\lambda,\mu)}$ and for any $x \in X$

$$y \not\leq_D f(x).$$

Proposition 3.4 (Nakayama[7])

Suppose that G is closed, and that there is at least a properly efficient solution to the primal problem. Then, under the condition of Slater's constraint qualification,

$$(Y_G)^c \subset \bigcup_{\substack{\mu \in \text{int } D^o \\ \lambda \in Q^o}} Y_{H^-(\lambda,\mu)} \subset \bigcup_{U \in \mathcal{U}} Y_{S(U)} \subset \text{cl } (Y_G)^c.$$

Lemma 3.4 (Nakayama[7])

The following holds:

$$\mathrm{Min}_D\,(P) = \mathrm{Min}_D\, Y_G.$$

Proposition 3.5

Assume that G is closed, that there exists at least a D-minimal solution to the primal problem, and that these solutions are all proper. Then, under the condition of Slater's constraint qualification, the following holds:

(i) $\mathrm{Min}_D\,(P) = \mathrm{Max}_D\,(D_N)$

(ii) $\mathrm{Min}_D\,(P) = \mathrm{Max}_D\,(D_J)$.

In some cases, one might not so much as expect that the G is closed. In this situation, we can invoke to some apropriate normality condition in order to derive the duality. In more detail, see for examle, Jahn[5], Borwein and Nieuwenhuis[1], and Sawaragi, Nakayama and Tanino[10]. In lenear cases, fortunately, it is readily seen that the set G is closed. In addition, we have $G = \mathrm{epi}\, W$, if there exists no $x \in M$ such that $(C-UA)x \leq_D 0$. Therefore, we can derive Isermann's duality in linear cases via the stated geometric duality. We shall discuss this in the following section.

4. Geometric Approach to Isermann's Duality in Linear Cases

Proposition 4.1

Let $f(x)=Cx$, $g(x)=Ax-b$ and $X'=M$, where C and A are $p{\times}m$

and $m \times n$ matrices, respectively and M is a pointed closed convex cone in R^n. Then every supporting hyperplane, $H(\lambda,\mu:\gamma)$ $(\gamma=\langle\mu,\hat{y}\rangle+\langle\lambda,\hat{z}\rangle)$, for epi W at an arbitrary point $(\hat{z},\hat{y})$ such that $\hat{y}$ $W(\hat{z})$, passes through the point $(z,y)=(b,0)$ independently of $(\hat{z},\hat{y})$. In addition, we have $\mu \in \text{int } D^o$, $\lambda \in Q^o$ and

$$C^T\mu - A^T\lambda \geqq_{M^o} 0. \tag{4.1}$$

Conversely, if $\mu \in D^o$ and $\lambda \in Q^o$ satisfy the relation (4.1), then the hyperplane with the normal (λ,μ) passing through the point $(z,y)=(b,0)$ supports epi W.

Proof: It has been shown in Lemma 3.2 that if the hyperplane $H(\lambda,\mu:\gamma)$ supports epi W, then $\mu \in D^o$ and $\lambda \in Q^o$. Further, since every efficient solution for linear cases is proper (See, for example, Sawaragi, Nakayama and Tanino[10]), we have $\mu \in \text{int } D^o$. Now, note that since $\hat{y} \in W(\hat{z})$, there exists $\hat{x} \in R^p$ such that

$$C\hat{x} = \hat{y}$$

$$b - A\hat{x} \leqq_Q \hat{z}.$$

Therefore, it follows from the supporting property of the hyperplane $H(\lambda,\mu:\gamma)$ that for any $(z,y) \in \text{epi } W$

$$\langle\mu,y\rangle + \langle\lambda,z\rangle \geqq \langle\mu,\hat{y}\rangle + \langle\lambda,\hat{z}\rangle$$

$$\geqq \langle\mu,C\hat{x}\rangle + \langle\lambda, b-A\hat{x}\rangle, \tag{4.2}$$

where the last half part of (4. 2) follows from the fact that $\lambda \in Q^o$ and $\hat{z} - (b- A\hat{x}) \in Q$. Since $(b-Ax, Cx) \in \text{epi } W$ for any $x \in M$, the relation (4.2) yields that for any $x \in M$

$$\langle\mu, Cx\rangle + \langle\lambda, b-Ax\rangle \geqq \langle\mu, C\hat{x}\rangle + \langle\lambda, b-A\hat{x}\rangle.$$

Consequently, for any $x \in M$

$$\langle C^T\mu-A^T\lambda, x-\hat{x}\rangle \geqq 0$$

and hence for any $x-\hat{x} \in M$

$$\langle C^T\mu - A^T\lambda,\ x-\hat{x}\rangle \geqq 0$$

Thereofore,

$$C^T\mu - A^T\lambda \geqq_{M^o} 0. \tag{4.3}$$

Seeing that the point $(b,0)$, which corresponds to $x=0$, belongs to epi W, it follows from (4.2) and (4.3) that

$$\langle \mu,\ y\rangle + \langle \lambda,\ \hat{z}\rangle = \langle \lambda,\ b\rangle.$$

This means that the supporting hyperplane $H(\lambda,\mu:\gamma)$ passes through the point $(z,y)=(b,0)$ independently of the given supporting point $(\hat{z},\hat{y})$.

Conversely, suppose that $\mu \in D^o$ and $\lambda \in Q^o$ satisfy the relation (4.1). Recall that for every $(z,y) \in$ epi W there exists $x \in M$, which may depend on (z,y), such that

$$y \in Cx + D \quad \text{and} \quad z - (b - Ax) \in Q.$$

It follows, therefore, that for any $\mu \in D^o$ and $\lambda \in D^o$

$$\langle \mu,\ y-Cx\rangle \geqq 0 \quad \text{and} \quad \langle \lambda,\ z-b+Ax\rangle \geqq 0. \tag{4.4}$$

Hence, by using the relation (4.1), we have from (4.4)

$$\langle \mu, y\rangle + \langle \lambda, z\rangle \geqq \langle \lambda, b\rangle \tag{4.5}$$

for every $(z,y) \in$ epi W. The realtion (4.5) shows that the hyperplane $H(\lambda,\mu:\gamma)$ passing through the point $(b,0)$ and satisfying $C^T\mu \geqq_{M^o} A^T\lambda$ supports epi W. This completes the proof.

The following lemma is an extension of the well known Stiemke's theorem and provides a key to clarify a relationship between Isermann's formulation and our geometric approach.

Lemma 4.1

There exists some $\mu \in \text{int } D^o$ such that

$$(C-UA)^T\mu \geqq_{M^o} 0 \tag{4.6}$$

if and only if there exists no $x \in M$ such that

$$(C-UA)x \leq_D 0. \tag{4.7}$$

Proof: Suppose first that there exists some $\mu \in \text{int } D^o$ such that (4.6) holds. If some $x \in M$ satisfy (4.7), or equivalently,

$$(C-UA)x \in (-D)\setminus\{0\}$$

then since $\mu \in \text{int } D^o$

$$\langle \mu, (C-UA)x \rangle < 0$$

which contradicts (4.6). Therefore, there is no $x \in M$ such that (4.7) holds.

Conversely, suppose that there exists no $x \in M$ such that (4.7) holds. This means

$$(C-UA)M \cap (-D) = \{0\},$$

from which we have

$$((C-UA)M)^o + (-D)^o = R^n.$$

Hence for an arbitrary $\mu_0 \in \text{int } D^o$ there exists $\mu_1 \in ((C-UA)M)^o$ and $\mu_2 \in (-D)^o$ such that

$$\mu_0 = \mu_1 + \mu_2 \tag{4.8}$$

and thus

$$\mu_1 = -\mu_2 + \mu_0.$$

Since $-\mu_2 \in D^o$ and $\mu_0 \in \text{int } D^o$, it follows from (4.8) that we have $\mu_1 \in ((C-UA)M)^o \cap \text{int } D^o$. Consequently, recalling that $((C-UA)M)^o = \{\mu | (C-UA)^T\mu \geqq_M o 0\}$, the existence of $\mu \in \text{int } D^o$ satisfying (4.6) is established. This completes the proof.

Proposition 4.2

For linear cases with $b \neq 0$,

$$\bigcup_{U \in \mathcal{U}_0} \{Ub\} = \bigcup_{U \in \mathcal{U}_0} \Phi(U) = \bigcup_{\substack{\lambda \in Q^o \\ \mu \in \text{int } D^o}} Y_{H(\lambda,\mu)}$$

Proof: According to Proposition 4.1 with $f(x)=Cx$ and $g(x)=Ax-b$, for $\mu \in \text{int } D^o$ and $\lambda \in Q^o$ such that $C^T\mu >_{=M^o} A^T\lambda$, we have

$$\langle \mu, f(x) \rangle + \langle \lambda, g(x) \rangle \geq \langle \lambda, b \rangle \quad \text{for all } x \in M.$$

Therefore, for $U \in R^{p \times m}$ such that $U^T\mu=\lambda$

$$\langle \mu, f(x)+Ug(x) \rangle \geq \langle \mu, Ub \rangle \text{ for all } x \in M,$$

which implies by virtue of the well known scalarization property and $\mu \in \text{int } D^o$ that

$$f(x) + Ug(x) \not\leq_D Ub \quad \text{for all } x \in X', \tag{4.9}$$

Hence for $U \in \mathcal{U}_0$

$$Ub \in \Phi(U),$$

which leads to $\bigcup_{U \in \mathcal{U}_0} \{Ub\} \subset \bigcup_{U \in \mathcal{U}_0} \Phi(U)$.

Next in order to show $\bigcup_{U \in \mathcal{U}_0} \Phi(U) \subset \bigcup_{\substack{\lambda \in Q^o \\ \mu \in \text{int } D^o}} Y_{H(\lambda,\mu)}$, suppose that $\bar{y} \in \Phi(U)$ for some $U \in \mathcal{U}_0$. Suppose further that $U^T\mu=\lambda$ and $C^T\mu \geq_{M^o} A^T\lambda$ for some $\mu \in \text{int } D^o$ and some $\lambda \in Q^o$. Then since from Lemma 4.1 we have $(C-UA)x \not\leq_D 0$ for all $x \in M$, we can guarantee the existence of an efficient solution $\bar{x} \in M$ for the vector valued Lagrangian $L(x,U)=Cx+U(b-Ax)$ such that $\bar{y}=C\bar{x}+U(b-A\bar{x})$. Moreover, since $L(.,U)$ is a convex vector-valued function over M for each U, due to the efficiency of $\bar{x}$ for $L(x,U)$ there exists $\bar{\mu} \in \mathrm{int}\, D^o$ such that

$$\langle \bar{\mu},\ C\bar{x}+U(b-A\bar{x})\rangle \leqq \langle \bar{\mu},\ Cx+U(b-Ax)\rangle \quad \text{for all } x \in M. \tag{4.10}$$

Hence, letting $\bar{\lambda}=U^T\bar{\mu}$

$$\langle \bar{\mu},\ \bar{y}\rangle \leqq \langle \bar{\mu},\ y\rangle + \langle \bar{\lambda},\ z\rangle \quad \text{for all } (z,y) \in \mathrm{epi}\, W. \tag{4.11}$$

which implies that $\bar{y} \in Y_{H(\lambda,\mu)}$. This establishes the desired inclusion.

Finally, we shall show $\bigcup_{\substack{\mu\in \mathrm{int}\, D^o \\ \lambda \in D^o}} Y_{H(\lambda,\mu)} \subset \bigcup_{U\in \mathcal{U}_0} \{Ub\}$. Suppose now that $\bar{y} \in Y_{H(\lambda,\mu)}$ for some $\mu \in \mathrm{int}\, D^o$ and $\lambda \in Q^o$. Since $(b,0)$ is a supporting point of $H(\lambda,\mu)$ for epi W according to Proposition 4.1, we have

$$\langle \mu, f(x)\rangle + \langle \lambda, g(x)\rangle \geqq \langle \lambda, b\rangle \quad \text{for all } x \in X' \tag{4.12}$$

and

$$\langle \mu, \bar{y}\rangle = \langle \lambda, b\rangle \tag{4.13}$$

Since $b \neq 0$, recall that the relation (4.13) shows that two equations $U^T\mu=\lambda$ and $Ub=\bar{y}$ have a common solution $U \in R^{p\times m}$ (Penrose[7]). In other words, we have $\bar{y} = Ub$ for some $U \in R^{p\times m}$ such that $U^T\mu=\lambda$, which leads to $\bar{y} \in \bigcup_{U\in \mathcal{U}_0} \{Ub\}$. This establishes the desired inclusion.

Now we can obtain the Isermann duality for linear vector cases via Propositions 3.4-3.5 and Lemma 3.4:

Theorem 4.1

For $b \neq 0$,

$$\mathrm{Min}_D\, (P_I) = \mathrm{Max}_D\, (D_I).$$

REFERENCES

[1] Borwein, J.M. and J.W. Nieuwenhuis, 'Two kinds of normality in vector optimization', Mathematical Programming 28, 185-191 (1984)

[2] Corley, H.W., 'Duality for maximizations with respect to cones', Journal of Mathematical Analysis and Applications 84, 560-568 (1981)

[3] Gale, D., The Theory of Linear Economic Models, McGraw Hill, NY, (1960)

[4] Isermann, H., 'On some relations between a dual pair of multiple objective linear programs', Zeitschrift fur Operations Research 22, 33-41 (1978)

[5] Jahn, J., 'Duality in vector optimization', Mathematical Programming 25, 343-353 (1983)

[6] Luenberger, D.G., Optimization by Vector Space Methods, Wiley, New York (1969)

[7] Nakayama, H., 'A geometric consideration on duality in vector optimization, Journal of Optimization Theory and Applications, 44, 625-65, (1984)

[8] Penrose, R, 'A Generalized Inverse for Matrices' Proceedings of Cambridge Philosophy and Society, Vol. 51, 406-413, (1955)

[9] Ritter, K., 'Optimization theory in linear spaces III', Mathematische Annalen 184, 133-154 (1970)

[10] Sawaragi, Y., H. Nakayama and T. Tanino, Theory of Multiobjective Optimization, Academic Press, to appear

[11] Tanino, T., and Y. Sawaragi, 'Duality theory in multiobjective programming', Journal of Optimization Theory and Applications 27, 509-529, (1979)

CONJUGATE MAPS AND CONJUGATE DUALITY

Tetsuzo Tanino
Department of Mechanical Engineering II
Tohoku University

Abstract

This paper considers conjugate duality in multiobjective optimization, in which minimality (efficiency, noninferiority, or Pareto optimality) is a natural solution concept. First, conjugate maps and subgradients are defined for vector-valued functions and point-to-set maps. Embedding the primal problem into a family of perturbed problems enables us to define a dual problem in terms of the conjugate map of the perturbed objective function. Every solution to the stable primal problem is associated with a solution to the dual problem, which is characterized as a subgradient of the perturbation map. This pair of solutions is also a saddle point of the Lagrangian.

1. Introduction and Preliminaries

This paper deals with the conjugate duality in multiobjective optimization (vector optimization). Some well-known concepts in ordinary conjugate duality, such as conjugate functions, subgradients and stability (see, e.g., Rockafellar[1,2]), will be extended to the case of multiobjective optimization.

In this paper we consider a multiobjective optimization problem

$$\underset{x}{\text{minimize}} \quad f(x) = \begin{pmatrix} f_1(x) \\ \vdots \\ f_p(x) \end{pmatrix},$$

where each f_i $(i=1,\dots,p)$ is a proper extended real-valued function from the n-dimensional Euclidean space R^n into $R \cup \{+\infty\}$. If some $f_i(x) = +\infty$, x is regarded to be infeasible. Hence we assume that $\text{dom } f_i = \{x \in R^n: f_i(x) < +\infty\} \neq \emptyset$ is the same for all $i=1,\dots,p$ and denote it by dom f. In this problem, our aim is to find minimal (efficient, nondominated or noninferior) solutions. The nonnegative orthant of R^p is denoted by R^p_+:

$$R^p_+ = \{y \in R^p : y \geqq 0^\dagger\}.$$

Definition 1 (Minimal, maximal). Given a set $Y \subset R^p$, a point $\hat{y} \in Y$ is said to be minimal (resp. maximal) in Y if

$$(Y - \hat{y}) \cap (-R^p_+) = \{0\} \quad (\text{resp. } (Y - \hat{y}) \cap R^p_+ = \{0\}),$$

i.e., if there is no $y \in Y$ such that $y \leq \hat{y}$ (resp. $y \geq \hat{y}$). The set of

---

† For $y, y' \in R^p$, $y \leqq y'$ iff $y_i \leqq y'_i$ for all $i=1,\dots,p$;
$y \leq y'$ iff $y \leqq y'$ and $y \neq y'$;
$y < y'$ iff $y_i < y'_i$ for all $i=1,\dots,p$.

all minimal (resp. maximal) points in Y is denoted by Min Y (resp. Max Y).

When p=1, Min Y and Max Y coincide with the sets {min Y} and {max Y} respectively. A point $\hat{x} \in R^n$ such that $f(\hat{x}) \in \text{Min}\ \{f(x) : x \in R^n\}$ is called a solution to the multiobjective optimization problem

$$\text{minimize}\quad f(x).$$

The following concepts are often useful in this paper.

Definition 2 (Min-complete, max-complete). A set $Y \subset R^p$ is said to be min-complete (resp. max-complete) if

$$Y \subset \text{Min}\ Y + R^p_+ \quad (\text{resp.}\quad Y \subset \text{Max}\ Y - R^p_+).$$

Definition 3 (R^p_+-closed, R^p_+-bounded). A set $Y \subset R^p$ is said to be R^p_+-closed if the set $Y + R^p_+$ is closed. Y is said to be R^p_+-bounded if

$$Y^+ \cap (- R^p_+) = \{0\},$$

where Y^+ is the extended recession cone (asymptotic cone) of Y defined by

$$Y^+ = \{y' \in R^p : \text{there exist sequences } \{\alpha_k\} \subset R \text{ and } \{y^k\} \subset Y \text{ such that } \alpha_k > 0,\ \alpha_k \to 0 \text{ and } \alpha_k y^k \to y'\}.$$

Lemma 1. If a set $Y \subset R^p$ is R^p_+-bounded and R^p_+-closed, then Y is min-complete.

Outline of proof: For an arbitrary $y \in Y$, let

$$Y' = (Y + R^p_+) \cap (y - R^p_+).$$

Since Y is R^p_+-bounded and R^p_+-closed, we can prove that Y' is compact. Hence, for any $\mu \in \mathring{R}^p_+ = \{\mu \in R^p : \mu > 0\}$, there exists $\hat{y} \in Y'$ such that

$$\langle\mu,\hat{y}\rangle = \min\ \{\langle\mu,y'\rangle : y' \in Y'\} \quad (\langle\cdot,\cdot\rangle \text{ is the inner product}).$$

Then $\hat{y} \in \text{Min}\ Y' \subset \text{Min}\ Y$. Therefore $Y \subset \text{Min}\ Y + R^p_+$.

2. Conjugate Maps

In this section, we introduce the concept of conjugate maps for vector-valued functions and point-to-set maps. When f is an extended real-valued function on R^n, i.e., when $f : R^n \to \bar{R} = R \cup \{\pm\infty\}$, its conjugate function f^* is defined by

$$f^*(t) = \sup \{\langle t,x\rangle - f(x) : x \in R^n\} \quad \text{for } t \in R^n.$$

Unfortunately, the concept of supremum of a set Y in R^p or $\bar{R}^p$ in the sense of maximality (efficiency) has not yet been established, though some definitions have been proposed (see Gros[3], Nieuwenhuis[4], Kawasaki[5,6], and Ponstein[7,8]. Note that the definition by Zowe[9] or Brumelle[10] is entirely different since it is given as a least upper bound with respect to the order relation $\leqq$ in R^p). Hence we use "Max" and "Min" as an extension of usual "sup" and "inf" respectively.

In order to consider the vector-valued case, we must define a paired space of R^n with respect to R^p. A most natural paired space is the set of all p×n matrices, which is denoted by $R^{p\times n}$. However, its dimension p×n is often too large. Another idea is to take $(R^n)^* = R^n$ as a paired space as in the scalar case and to consider the p-dimensional vector $\langle\langle t,x\rangle\rangle := (\langle t,x\rangle,\ldots,\langle t,x\rangle)^T$ for $t \in R^n$, where superscript T denotes the transpose. Since $\langle\langle t,x\rangle\rangle = (t,\ldots,t)^T x$, the vector variable is a special case of matrix variable. Hence we mainly discuss the case of matrix variable, though many results in this paper are valid for both cases.

Definition 4 (Conjugate map , biconjugate map). Let F be a point-to-set map from R^n into R^p. The point-to-set map $F^*: R^n \to R^p$, defined by

$$F^*(T) = \text{Max} \bigcup_{x \in R^n} [Tx - F(x)] \quad \text{for } T \in R^{p\times n}$$

is called the conjugate map of F. Moreover, the conjugate map F^{**} of F^* defined by

$$F^{**}(x) = \text{Max} \bigcup_{T \in R^{p\times n}} [Tx - F^*(T)]$$

is called the biconjugate map of F. When f is a vector-valued function from R^n to $R^p \cup \{+\infty\}$, let dom f = $\{x \in R^n : f(x) \neq +\infty\}$ and define the conjugate map f^* of f by

$$f^*(T) = \text{Max} \{Tx - f(x) : x \in \text{dom } f\}.$$

Here, $+\infty$ is the imaginary point whose every component is $+\infty$. Namely we identify the function f with the point-to-set map which is equal to $\{f(x)\}$ for $x \in$ dom f and is empty otherwise. The biconjugate map f^{**} can be defined as the conjugate map of f^*.

<u>Proposition 1</u>. Let F be a point-to-set map from R^n into R^p and $\bar{x} \in R^n$. If we define another point-to-set map G by $G(x) = F(x + \bar{x})$, then

(i) $G^*(T) = F^*(T) - T\bar{x}$;

(ii) $G^{**}(x) = F^{**}(x + \bar{x})$.

Proof: Easy.

<u>Proposition 2</u>. Let F be a point-to-set map from R^n into R^p and $\bar{y} \in R^p$. Then

(i) $(F + \bar{y})^*(T) = F^*(T) - \bar{y}$;

(ii) $(F + \bar{y})^{**}(x) = F^{**}(x) + \bar{y}$,

where $F + \bar{y}$ is a point-to-set map defined by $(F + \bar{y})(x) = F(x) + \bar{y}$.

Proof: Easy.

Proposition 3 (An extension of Fenchel's inequality). Let F be a point-to-set map from R^n into R^p. If $y \in F(\hat{x})$ and $y' \in F^*(T)$, then

$$y + y' \nleq T\hat{x} \quad (\text{i.e. } T\hat{x} - (y + y') \notin R^p_+ \setminus \{0\}).$$

Proof: Since $y \in F(\hat{x})$, $T\hat{x} - y \in \bigcup_x [Tx - F(x)]$. Hence, if $y' \leq T\hat{x} - y$, then it contradicts the assumption $y' \in F^*(T) = \text{Max} \bigcup_x [Tx - F(x)]$.

Corollary 1. Let F be a point-to-set map from R^n into R^p. If $y \in F(0)$ and $y' \in -F^*(T)$, then $y \nleq y'$. Moreover, if $y \in F(x)$ and $y' \in F^{**}(x)$, then $y \nleq y'$.

Proof: Immediate.

Lemma 2. Let F_1 and F_2 be point-to-set maps from R^n into R^p. Then

$$\text{Max} \bigcup_x [F_1(x) + F_2(x)] \subset \text{Max} \bigcup_x [F_1(x) + \text{Max } F_2(x)].$$

If $F_2(x)$ is max-complete (i.e. $F_2(x) \subset \text{Max } F_2(x) - R^p_+$) for every $x \in R^n$, then the converse inclusion also holds.

Proof: Let $\hat{y} \in \text{Max} \bigcup_x [F_1(x) + F_2(x)]$. Then there exists $\hat{x} \in R^n$ such that $y = y^1 + y^2$ for some $y^1 \in F_1(\hat{x})$ and $y^2 \in F_2(\hat{x})$. If we suppose that $y^2 \notin \text{Max } F_2(\hat{x})$, there exists $\bar{y}^2 \in F_2(\hat{x})$ such that $y^2 \leq \bar{y}^2$. Then $y = y^1 + y^2 \leq y^1 + \bar{y}^2$, which is a contradiction. Therefore $y^2 \in \text{Max } F_2(\hat{x})$. Since $\bigcup_x [F_1(x) + F_2(x)] \supset \bigcup_x [F_1(x) + \text{Max } F_2(x)]$, $\hat{y} \in \text{Max} \bigcup_x [F_1(x) + \text{Max } F_2(x)]$.

Next, suppose that $F_2(x)$ is max-complete for every x, then

$$F_2(x) - R^p_+ = \text{Max } F_2(x) - R^p_+$$

for every x. Thus

$$F_1(x) + F_2(x) - R^p_+ = F_1(x) + \text{Max } F_2(x) - R^p_+$$

for every x and so

$$\bigcup_x [F_1(x) + F_2(x)] - R^p_+ = \bigcup_x [F_1(x) + \text{Max } F_2(x)] - R^p_+ .$$

Taking the Max of both sides, we have

$$\text{Max} \bigcup_x [F_1(x) + F_2(x)] = \text{Max} \bigcup_x [F_1(x) + \text{Max } F_2(x)],$$

since $\text{Max } Y = \text{Max } (Y - R^p_+)$ for any $Y \subset R^p$. This completes the proof.

Corollary 2. Let F be a point-to-set map from R^n into R^p. Then

$$F^*(T) \subset \text{Max} \bigcup_x [Tx - \text{Min } F(x)].$$

Moreover, if F(x) is min-complete for every x, then the equliaty holds.

Proof: Take $F_1(x) = \{Tx\}$ and $F_2(x) = -F(x)$ in Lemma 2.

Corollary 3. Let F be a point-to-set map from R^n into R^p. If F(x) is max-complete for every x, then

$$\text{Max} \bigcup_x F(x) = \text{Max} \bigcup_x \text{Max } F(x).$$

Proof: Take $F_1(x) = \{0\}$ and $F_2(x) = F(x)$ in Lemma 2.

Remark 1. In the above corollaries and lemma, we cannot dispense with the completeness condition of F(x) for the equality. For example, let $F : R \to R^2$ be

$$F(x) = \begin{cases} \begin{pmatrix} 0 \\ 0 \end{pmatrix} & \text{if } x = 0 \\ \{y \in R^2 : y_1^2 + y_2^2 < 1\} & \text{if } x \neq 0. \end{cases}$$

Then, for $T = (0,0)^T$,

$$F^*(T) = \emptyset ,$$

while

$$\operatorname{Max} \bigcup_x [Tx - \operatorname{Min} F(x)] = \{0\}.$$

Similarly, $\operatorname{Max} \bigcup_x F(x) = \emptyset$, but $\operatorname{Max} \bigcup_x \operatorname{Max} F(x) = \{0\}$. This drawback is probably due to the fact that we are using "Max" instead of "Sup".

3. Subgradients of Vector-Valued Functions and Point-to-Set Maps

In this section, we will introduce the concept of subgradients of vector-valued functions and point-to-set maps. A subgradient t of a scalar-valued function f at $\hat{x}$ is defined by

$$f(x) \geqq f(\hat{x}) + \langle t, x-\hat{x}\rangle \quad \text{for any } x \in R^n.$$

The definition can be formally extended to nonconvex functions, though it essentially requires (at least local) convexity. A subgradient has the geometrical interpretation as the normal vector to a supporting hyperplane of epi f at $(\hat{x}, f(\hat{x}))$. Thus we may extend the definition of subgradients to the vector-valued case as follows.

Definition 5 (Subgradient, subdifferential). (i) Let f be a function from R^n to $R \cup \{+\infty\}$. A $p \times n$ matrix T is said to be a subgradient of f at $\hat{x} \in \operatorname{dom} f$ if

$$f(x) \nleq f(\hat{x}) + T(x - \hat{x}) \quad \text{for any } x \in R^n,$$

i.e., if

$$f(\hat{x}) - T\hat{x} \in \operatorname{Min} \{f(x) - Tx \in R^p : x \in R^n\}$$
$$= \operatorname{Min} \{f(x) - Tx : x \in \operatorname{dom} f\}.$$

The set of all subgradients of f at $\hat{x}$ is called the subdifferential of f at $\hat{x}$ and is denoted by $\partial f(\hat{x})$. If $\partial f(\hat{x})$ is not empty, f is said to be subdifferentiable at $\hat{x}$.

(ii) Let F be a point-to-set map from R^n into R^p and $\hat{y} \in F(\hat{x})$. A p×n matrix T is said to be a subgradient of F at $(\hat{x};\hat{y})$ if

$$\hat{y} - T\hat{x} \in \text{Min} \bigcup_{x \in R^n} [F(x) - Tx].$$

The set of all subgradients of F at $(\hat{x};\hat{y})$ is called the subdifferential of F at $(\hat{x};\hat{y})$ and is denoted by $\partial F(\hat{x};\hat{y})$. If $\partial F(\hat{x};\hat{y})$ is not empty for every $\hat{y} \in F(\hat{x})$, F is said to be subdifferentiable at $\hat{x}$.

Remark 2. It is clear that $F(\hat{x}) = \text{Min } F(\hat{x})$ if F is subdifferentiable at $\hat{x}$.

Remark 3. T is a subgradient of F at $(\hat{x};\hat{y})$ if and only if the polyhedral convex set

$$\{(x,y) : y \leqq \hat{y} + T(x - \hat{x})\}$$
$$= \{(x,y) : \langle (t^i,-e^i),(x,y)\rangle \geqq \langle (t^i,-e^i),(\hat{x},\hat{y})\rangle,\ i=1,\dots,p\}$$

is a supporting polyhedral convex set of epi F at $(\hat{x},\hat{y})$, where t^i is the i-th row vector of T, e^i is the i-th unit vector and

$$\text{epi } F = \{(x,y) \in R^n \times R^p : y \in F(x) + R^p_+\}.$$

Thus F needs not to be convex in the sense that epi F is a convex set in order for F to be subdifferentiable. For example, let F be a constant point-to-set map on R :

$$F(x) = \{y \in R^2 : y_1 y_2 = 1,\ y_1 < 0\}.$$

Then 0 matrix is a subgradient of F at (x;y) for any x and $y \in F(x)$. On the other hand, epi F is not convex.

Remark 4. Unlike the scalar case, the subdifferential is not necessarily a closed convex set even when f is a finite convex vector-valued function.

(i) Let $f(x) = \begin{pmatrix} x \\ -x \end{pmatrix}$. Then $T = \begin{pmatrix} 0 \\ t \end{pmatrix} \in \partial f(0)$ for any $t > -1$. However, $\begin{pmatrix} 0 \\ -1 \end{pmatrix} \notin \partial f(0)$. Thus $\partial f(0)$ is not closed.

(ii) Let $f(x) = \begin{pmatrix} x^2 \\ x^2 \end{pmatrix}$ for $x \in R$. Then we can easily check that $\begin{pmatrix} 2 \\ 0 \end{pmatrix}, \begin{pmatrix} 0 \\ 2 \end{pmatrix} \in \partial f(0)$, but $\begin{pmatrix} 1 \\ 1 \end{pmatrix} \notin \partial f(0)$. Thus $\partial f(0)$ is not convex.

The following proposition provides a characterization of minimal solutions by the subgradient 0, that is, the stationarity condition in an extended sense.

Proposition 4. (i) Let f be a function from R^n to $R^p \cup \{+\infty\}$. Then $f(\hat{x}) \in \text{Min}\,\{f(x) : x \in \text{dom}\, f\}$ if and only if $0 \in \partial f(\hat{x})$.

(ii) Let F be a point-to-set map from R^n into R^p. For $\hat{y} \in F(\hat{x})$, $\hat{y} \in \text{Min} \bigcup_x F(x)$ if and only if $0 \in \partial F(\hat{x};\hat{y})$.

Proof: Immediate from the definitions of subgradients.

The following propositions show the relationships between conjugate or biconjugate maps and subgradients.

Proposition 5. (i) Let f be a function from R^n to $R^p \cup \{+\infty\}$. Then $T \in \partial f(x)$ if and only if $Tx - f(x) \in f^*(T)$.

(ii) Let F be a point-to-set map from R^n into R^p. Then, for $y \in F(x)$, $T \in \partial F(x;y)$ if and only if $Tx - y \in F^*(T)$.

Proof: Obvious from the definitions of conjugate maps and

subgradients.

Proposition 6. (i) Let f be a function from R^n to $R^p \cup \{+\infty\}$. Then, $f(\hat{x}) \in f^{**}(\hat{x})$ if and only if $\partial f(\hat{x}) \neq \emptyset$, i.e., if and only if f is subdifferentiable at $\hat{x}$.

(ii) Let F be a point-to-set map from R^n into R^p. For $\hat{y} \in F(\hat{x})$, $\partial F(\hat{x};\hat{y}) \neq \emptyset$ if and only if $\hat{y} \in F^{**}(\hat{x})$. Thus F is subdifferentiable at $\hat{x}$ if and only if $F(\hat{x}) \subset F^{**}(\hat{x})$.

Proof: We prove (ii), because the proof of (i) is similar. From Proposition 5, $\partial F(\hat{x};\hat{y}) \neq \emptyset$ if and only if there exists $T \in R^{p\times n}$ such that

$$T\hat{x} - \hat{y} \in F^*(T).$$

Hence, if $\hat{y} \in F^{**}(\hat{x})$, then it is clear that $\partial F(\hat{x};\hat{y}) \neq \emptyset$. Conversely, suppose that $\partial F(\hat{x};\hat{y}) \neq \emptyset$, namely that $\hat{y} \in T\hat{x} - F^*(T)$ for some T. From Proposition 3, we have $\hat{y} \not\leq y$ for any $y \in \bigcup_{T'} [T'x - F^*(T')]$. Therefore $\hat{y} \in F^{**}(\hat{x})$.

As pointed out in Remark 3, convexity of F is not always necessary for its subdifferentiability. It is, however, often sufficient as shown in the next proposition.

Proposition 7. Let F be a convex point-to-set map from R^n into R^p satisfying $\text{Min}\, F(\hat{x}) = F(\hat{x})$. Suppose that $F(x) \neq \emptyset$ in some neighborhood of $\hat{x}$ and each $F^*(T)$ can be characterized completely by scalarization, i.e.

$$-F^*(T) = \{\hat{y} \in \bigcup_x [F(x) - Tx] : \langle\mu,\hat{y}\rangle = \min\{\langle\mu,y-Tx\rangle : y \in F(x),\ x \in R^n\} \quad \text{for some } \mu \in R^p_+\backslash\{0\}\}.$$

Then F is subdifferentiable at $\hat{x}$.

Proof: We may assume without loss of generality that $\hat{x} = 0$. Let $\hat{y} \in F(0) = \text{Min } F(0)$. Then $(0,\hat{y})$ is clearly a boundary point of epi F, which is a convex set from the assumption. Hence, by the support theorem of convex sets, there exists $(\lambda,\mu) \neq 0 \in R^n \times R^p$ such that

$$\langle\mu,\hat{y}\rangle \leqq \langle\lambda,x\rangle + \langle\mu,y\rangle \quad \text{for all } (x,y) \in \text{epi } F.$$

Since we can take every component of y as large as desired, $\mu \geqq 0$. If we assume that $\mu = 0$,

$$\langle\lambda,x\rangle \geqq 0 \quad \text{for all } (x,y) \in \text{epi } F.$$

Since $F(x) \neq \emptyset$ in some neighborhood of 0, this implies that $\lambda = 0$ and so leads to a contradiction. Hence $\mu \neq 0$ and we can choose $T \in R^{p\times n}$ as $-T^T\mu = \lambda$. Then

$$\langle\mu,\hat{y}\rangle \leqq \langle\mu,y - Tx\rangle \quad \text{for all } (x,y) \in \text{epi } F,$$

and so

$$\langle\mu,\hat{y}\rangle \leqq \langle\mu,y - Tx\rangle \quad \text{for all } x \in R^n \text{ and } y \in F(x).$$

Since $F^*(T)$ is assumed to be characterized completely by scalarization, the above inequality implies that $\hat{y} \in -F^*(T)$. Hence $T \in \partial F(0;\hat{y})$ from Proposition 5, and the proof is completed.

Remark 5. In view of the above proof, we may assume that $\mu \in S^p = \{\mu' \in R^p : \mu' \geqq 0, \sum_{i=1}^{p} \mu'_i = 1\}$, and therefore may take $T^T = (-\lambda,\ldots,-\lambda)$.

We have the following relationship between $f(x)$ and $f_i(x)$ $(i=1,\ldots,p)$ for convex vector-valued functions.

Proposition 8. Let f be a convex vector-valued function from R^n to

$R^p \cup \{+\infty\}$ such that ri (dom f) $\neq \emptyset$. Then

$$\partial f(\hat{x}) \subset \bigcup_{\mu \in S^p} \{T \in R^{p\times n} : T^T\mu \in \sum_{i=1}^{p} \mu_i \partial f_i(\hat{x})\}, \quad \text{for any } \hat{x},$$

where $S^p = \{\mu \in R^p : \mu \geqq 0, \sum_{i=1}^{p} \mu_i = 1\}$. Conversely, if $T^T\mu \in \sum_{i=1}^{p} \mu_i \partial f_i(\hat{x})$ for some $\mu \in S^p$ and if $-f^*(T)$ can be characterized completely by scalarization, then $T \in \partial f(\hat{x})$.

Proof: If $T \in \partial f(\hat{x})$, then from Proposition 5,

$$f(\hat{x}) - T\hat{x} \in -f^*(T) = \text{Min}\{f(x) - Tx : x \in \text{dom } f\}.$$

Since $f(x) - Tx$ is a convex function, there exists $\mu \in S^p$ such that

$$\langle \mu, f(\hat{x})\rangle - \langle T^T\mu, x\rangle = \min\{\langle \mu, f(x)\rangle - \langle T^T\mu, x\rangle : x \in \text{dom } f\}.$$

Hence,

$$0 \in \sum_{i=1}^{p} \mu_i \partial f_i(\hat{x}) - T^T\mu.$$

Conversely, if $T^T\mu \in \sum_{i=1}^{p} \mu_i \partial f_i(\hat{x})$ and if $f^*(T)$ can be characterized completely by scalarization, then we can trace the above procedure inversely.

<u>Proposition 9</u>. Let f be a vector-valued function from R^n to $R^p \cup \{+\infty\}$. If $t^i \in \partial f_i(\hat{x})$ for every $i=1,\ldots,p$, then $T = (t^1,\ldots,t^p)^T \in \partial f(\hat{x})$.

Proof: Obvious.

In convex analysis, it is well known that the biconjugate function f^{**} of a scalar-valued closed convex function f coincides with f itself. The following propositions provide extensions of this result to the vector-valued case. Since the proofs are rather long and tedious, they are omitted here. Details can be seen in Sawaragi et al[11].

<u>Proposition 10</u>. Let f be a convex function from R^n to $R^p \cup \{+\infty\}$. If each f_i $(i=1,\ldots,p)$ is subdifferentiable at $\hat{x}$ and if the set $\{f(x) - Tx \in R^p : x \in R^n\}$ is R^p_+-closed for every $T \in R^{p\times n}$, then

$$f^{**}(\hat{x}) = \{f(x)\}.$$

<u>Remark 6</u>. Note that if $\hat{x} \in \text{ri}\,(\text{dom}\, f)$, then every f_i is subdifferentiable at $\hat{x}$.

<u>Remark 7</u>. Proposition 6.10 is not necessarily valid for the vector dual variable. For example, let

$$f(x) = \begin{cases} \begin{pmatrix} x \\ x^2 \end{pmatrix} & \text{if } 0 \leqq x \leqq 1 \\ +\infty & \text{otherwise.} \end{cases}$$

Then $\begin{pmatrix} \frac{1}{8} \\ -\frac{1}{16} \end{pmatrix} \in [-f^*(\frac{1}{2})] \cap [\text{Max} \bigcup_t (-f^*(t))].$

<u>Proposition 11</u>. Let f be a closed† convex function from R^n to $R^p \cup \{+\infty\}$. If f is subdifferentiable at each $x \in \text{dom}\, f$ and if $f(\hat{x}) = +\infty$, then $f^{**}(\hat{x}) = \emptyset$.

† A vector-valued function f is said to be closed if its epigraph epi f is a closed set.

4. Duality in Multiobjective Optimization

In this section we derive duality results regarding a multiobjective optimization problem

$$\text{(P)} \qquad \underset{x}{\text{minimize}} \;\; f(x) \; ,$$

where f is an extended real vector-valued function from R^n to $R^p \cup \{+\infty\}$. In other words, (P) is the problem to find $\hat{x} \in R^n$ such that

$$f(\hat{x}) \in \text{Min} \{f(x) \in R^p : x \in R^n\}.$$

We denote the set $\text{Min} \{f(x) \in R^p : x \in R^n\}$ simply by Min (P) and call every $\hat{x}$ satisfying $f(\hat{x}) \in \text{Min (P)}$ a solution to the problem (P).

As in ordinary conjugate duality theory, we analyze the problem (P) by embedding it in a family of perturbed problems. The space of perturbation is assumed to be the Euclidean space R^m for simplicity. Thus we consider a function $\phi : R^n \times R^m \to R^p \cup \{+\infty\}$ such that

$$\phi(x,0) = f(x) \quad \text{for any } x \in R^n,$$

and a family of perturbed problems

$$(P_u) \qquad \underset{x}{\text{minimize}} \;\; \phi(x,u)$$

or

(P_u) Find $\hat{x} \in R^n$ such that

$$\phi(\hat{x},u) \in \text{Min} \{\phi(x,u) \in R^p : x \in R^p\}.$$

First, we define the perturbation map for (P) which is an extension of the perturbation function or the optimal value function w(u) in ordinary scalar optimization, which is defined by

$$w(u) = \inf \{\phi(x,u) : x \in R^n\}.$$

Let Y and W be point-to-set maps from R^m into R^p defined by

$$Y(u) = \{\phi(x,u) \in R^p : x \in R^n\}$$

and

$$W(u) = \text{Min } Y(u) = \text{Min } \{\phi(x,u) \in R^p : x \in R^n\}$$

respectively. Of course

$$\text{Min } (P) = W(0).$$

Lemma 3. If the function ϕ is convex on $R^n \times R^m$ and $Y(u)$ is min-complete (i.e. $Y(u) \subset W(u) + R^p_+$) for each $u \in R^m$, then the perturbation map W is convex.

Proof: If the assumptions of the lemma hold,

$$\begin{aligned} \text{epi } W &= \{(u,y) \in R^m \times R^p : y \in W(u) + R^p_+\} \\ &= \{(u,y) \in R^m \times R^p : y \in Y(u) + R^p_+\} \\ &= \{(u,y) \in R^m \times R^p : y \geqq \phi(x,u),\ x \in R^n\}. \end{aligned}$$

The last set is the image of epi ϕ under the projection $(x,u,y) \to (u,y)$. Since epi ϕ is convex and convexity is preserved under projection, epi W is convex.

In order to define the dual problem, we consider the conjugate map ϕ^* of ϕ, i.e.,

$$\phi^*(T,\Lambda) = \text{Max } \{Tx + \Lambda u - \phi(x,u) \in R^p : x \in R^n,\ u \in R^m\}.$$

Note that ϕ^* is not a function, but a point-to-set map from $R^{p\times n} \times R^{p\times m}$ into R^p. However, we consider the following problem as the dual problem of the problem (P) with respect to the given perturbations:

(D) Find $\hat{\Lambda} \in R^{p\times m}$ such that

$$- \phi^*(0,\hat{\Lambda}) \cap \text{Max} \bigcup_{\Lambda} (- \phi^*(0,\Lambda)) \neq \emptyset.$$

This problem may be written formally as

$$\text{(D)} \qquad \underset{\Lambda}{\text{maximize}} \quad -\phi^*(0,\Lambda).$$

However, it is not an ordinary multiobjective optimization problem, since $-\phi^*(0,\cdot)$ is not a function but a point-to-set map. The set $\text{Max} \bigcup_{\Lambda} (-\phi^*(0,\Lambda))$ will be simply denoted by Max (D) and every $\hat{\Lambda}$ satisfying the above relation will be called a solution to the problem (D).

The first result is the so-called weak duality theorem, which implies that each feasible value of the primal minimization problem (P) is not less (in the ordering $\geqq$) than any feasible value of the dual maximization problem (D).

Proposition 12. For any $x \in R^n$ and $\Lambda \in R^{p\times m}$,

$$f(x) \not\in -\phi^*(0,\Lambda) - R^p_+\backslash\{0\}.$$

Proof: Let $y = f(x) = \phi(x,0)$ and $y' \in \phi^*(0,\Lambda)$. Then, from Proposition 3,

$$y + y' \nleq 0x + \Lambda 0 = 0.$$

Corollary 4. For any $y \in \text{Min (P)}$ and $y' \in \text{Max (D)}$, $y \nleq y'$.

Proof: Immediate from Proposition 12.

We can prove that the conjugate map W^* of the perturbation map W is directly connected with ϕ^* as in the following lemma.

Lemma 4. The following relation holds for every $\Lambda \in R^{p\times m}$:

$$W^*(\Lambda) \supset \phi^*(0,\Lambda)$$

with the equality holding when Y(u) is min-complete for every $u \in R^m$.

Proof: $\phi^*(0,\Lambda) = \text{Max } \{0x + \Lambda u - \phi(x,u) \in R^p : x \in R^n, u \in R^m\}$

$$= \text{Max } \bigcup_u [\Lambda u - Y(u)].$$

Hence, in view of Corollary 2,

$$\phi^*(0,\Lambda) \subset \text{Max } \bigcup_u [\Lambda u - W(u)] = W^*(\Lambda)$$

with the equality holding when every Y(u) is min-complete. This completes the proof.

Throughout this paper we assume that Y(u) is min-complete for each $u \in R^m$. This assumption is satisfied fairly generally. Thus we can rewrite the dual problem (D) as follows:

(D) Find $\hat{\Lambda} \in R^{p\times m}$ such that

$$- W^*(\hat{\Lambda}) \cap \text{Max } \bigcup_\Lambda [- W^*(\Lambda)] \neq \emptyset.$$

<u>Lemma 5</u>.

$$\text{Max (D)} = \text{Max } \bigcup_\Lambda [- W^*(\Lambda)] = W^{**}(0).$$

Proof: Immediate from Lemma 4 and the definition of W^{**}.

Thus

$$\text{Min (P)} = W(0) \quad \text{and} \quad \text{Max (D)} = W^{**}(0).$$

Therefore the discussion on the duality (i.e. the relationship between Min (P) and Max (D) can be replaced by the discussion on the relationship between W(0) and $W^{**}(0)$. Proposition 6 justifies considering the following class of multionjective optimization problems.

<u>Definition 6</u> (Stable problem). The multiobjective optimization problem (P) is said to be stable if the perturbation map W is

subdifferentiable at 0.

Proposition 13. If the function ϕ is convex, $Y(u) \neq \emptyset$ for every u in some neighborhood of 0, and $W^*(\Lambda)$ can be characterized completely by scalarization for each Λ, then the problem (P) is stable. Here we should note that we may take a vector subgradient λ as pointed out in Remark 5.

Proof: From the assumptions, the point-to-set map Y is convex. Noting that $W(u) \subset Y(u)$ for every u, we can prove the proposition in a quite similar manner to the proof of Proposition 7.

In view of Proposition 6 (ii), the problem (P) is stable if and only if

$$\text{Min (P)} = W(0) \subset W^{**}(0) = \text{Max (D)}.$$

Thus we have the following theorem which is a generalization of the strong duality theorem in scalar optimization.

Theorem 1. (i) The problem (P) is stable if and only if, for each solution $\hat{x}$ to (P), there exists a solution $\hat{\Lambda}$ to the dual problem (D) such that

$$\phi(\hat{x},0) \in -\phi^*(0,\hat{\Lambda}), \quad \text{i.e.,} \quad f(\hat{x}) \in -W^*(\hat{\Lambda}),$$

or equivalently

$$(0,\hat{\Lambda}) \in \partial\phi(\hat{x},0), \quad \text{i.e.,} \quad \hat{\Lambda} \in \partial W(0;f(\hat{x})).$$

(ii) Conversely, if $\hat{x} \in R^n$ and $\hat{\Lambda} \in R^{p\times m}$ satisfy the above relation, then $\hat{x}$ is a solution to (P) and $\hat{\Lambda}$ is a solution to (D).

Proof: These results are obvious from the previous discussions.

5. Lagrangians and Saddle Points

In this section we define the Lagrangian and its saddle points for the problem (P) and investigate their properties.

Definition 7 (Lagrangian). The point-to-set map $L : R^n \times R^{p\times m} \to R^p$, defined by

$$- L(x,\Lambda) = \mathrm{Max}\ \{ \Lambda u - \phi(x,u) \in R^p : u \in R^m\},$$

i.e.

$$L(x,\Lambda) = \mathrm{Min}\ \{\phi(x,u) - \Lambda u \in R^p : u \in R^m\}$$

is called the Lagrangian of the multiobjective optimization problem (P) relative to the given perturbations.

We can write

$$L(x,\Lambda) = - \phi_x^*(\Lambda),$$

where ϕ_x denotes the function $u \to \phi(x,u)$ for a fixed $x \in R^n$ and ϕ_x^* denotes the conjugate map of ϕ_x.

Definition 8 (Saddle point). A point $(\hat{x},\hat{\Lambda}) \in R^n \times R^{p\times m}$ is called a saddle point of the Lagrangian L if

$$L(\hat{x},\hat{\Lambda}) \cap [\mathrm{Max} \bigcup_{\Lambda} L(\hat{x},\Lambda)] \cap [\mathrm{Min} \bigcup_{x} L(x,\hat{\Lambda})] \neq \emptyset.$$

It will be useful to express the problems (P) and (D) in terms of the Lagrangian L.

Proposition 14. Suppose that, for each $\Lambda \in R^{p\times m}$, either the set $\{\Lambda u - \phi(x,u) \in R^p : u \in R^m\}$ is max-complete for every $x \in R^n$, or it is $(- R^p_+)$-unbounded or empty for every $x \in R^n$, then

$$\phi^*(0,\Lambda) = \operatorname{Max} \bigcup_x [-L(x,\Lambda)]$$

and hence the dual problem (D) can be written formally as

$$\text{(D)} \qquad \operatorname{Max} \bigcup_\Lambda \operatorname{Min} \bigcup_x L(x,\Lambda).$$

Proof: In the former case, in view of Corollary 3,

$$\operatorname{Max} \bigcup_x [-L(x,\Lambda)] = \operatorname{Max} \bigcup_x \operatorname{Max} \{\Lambda u - \phi(x,u) \in R^p : u \in R^m\}$$

$$= \operatorname{Max} \bigcup_x \{\Lambda u - \phi(x,u) \in R^p : u \in R^m\}$$

$$= \operatorname{Max} \{0x + \Lambda u - \phi(x,u) \in R^p : x \in R^n,\ u \in R^m\}$$

$$= \phi^*(0,\Lambda).$$

On the other hand, in the latter case, both $\bigcup_x [-L(x,\Lambda)]$ and $\phi^*(0,\Lambda)$ are empty. Hence

$$\operatorname{Max}(D) = \operatorname{Max} \bigcup_\Lambda [-\phi^*(0,\Lambda)] = \operatorname{Max} \bigcup_\Lambda (-\operatorname{Max} \bigcup_x [-L(x,\Lambda)])$$

$$= \operatorname{Max} \bigcup_\Lambda [\operatorname{Min} \bigcup_x L(x,\Lambda)].$$

Since $L(x,\Lambda) = -\phi_x^*(\Lambda)$,

$$\operatorname{Max} \bigcup_\Lambda L(x,\Lambda) = \operatorname{Max} \bigcup_\Lambda [-\phi_x^*(\Lambda)] = \phi_x^{**}(0).$$

Hence we can directly apply Propositions 10 and 11 to obtain the following result.

<u>Proposition 15</u>. Suppose that ϕ_x is convex for each fixed $x \in R^n$, and that (i) when $f(x) = \phi(x,0)$ is finite, each $\phi_i(x,\cdot)$ is subdifferentiable at 0, and the set $\{\phi(x,u) - \Lambda u \in R^p : u \in R^m\}$ is R^p_+-closed for each $\Lambda \in R^{p\times m}$; (ii) when $f(x) = \phi(x,0) = +\infty$, ϕ_x is closed and

subdifferentiable at each $u \in \mathrm{dom}\ \phi_x$. Then, in case (i), $\mathrm{Max} \bigcup_{\Lambda} L(x,\Lambda) = \{\phi_x(0)\} = \{f(x)\}$, and in case (ii), $\mathrm{Max} \bigcup_{\Lambda} L(x,\Lambda) = \emptyset$. Thus the primal problem (P) can be written formally as

$$\text{(P)} \qquad \mathrm{Min} \bigcup_{x} \mathrm{Max} \bigcup_{\Lambda} L(x,\Lambda).$$

These expressions of the problems (P) and (D) suggest that the saddle point of the Lagrangian is closely connected with a pair of solutions to the problems (P) and (D).

Theorem 2. Suppose that the assumptions in Propositions 14 and 15 are both satisfied. Then the following statements are equivalent to each other:

(i) $(\hat{x},\hat{\Lambda})$ is a saddle point of L ;

(ii) $\hat{x}$ is a solution to the primal problem (P), $\hat{\Lambda}$ is a solution to the dual problem (D), and

$$\phi(\hat{x},0) \in -\phi^*(0,\hat{\Lambda}).$$

Proof: This theorem is obvious from Propositions 12, 14 and 15.

Theorem 3. Suppose that the hypothesis of Theorem 2 holds and that the problem (P) is stable. Then $\hat{x} \in R^n$ is a solution to (P) if and only if there exists $\hat{\Lambda} \in R^{p \times m}$ such that $(\hat{x},\hat{\Lambda})$ is a saddle point of the Lagrangian L.

Results similar to the above theorems for convex problems can be obtained with the vector dual variable (cf. Remark 5). The details can be seen in Tanino and Sawaragi[12].

6. Duality in Multiobjective Programming

In this section we apply the results obtained in the preceding sections to multiobjective programming.

We consider the following multiobjective programming problem:

$$\text{(P)} \qquad \text{minimize} \quad f(x) \quad \text{subject to} \quad g(x) \leqq 0 \,,$$

where f and g are vector-valued functions on R^n of dimension p and m, respectively. Of course we assume the existence of feasible solutions, i.e. let $X = \{x \in R^n : g(x) \leqq 0\} \neq \emptyset$. If we define a vector-valued function $\tilde{f} : R^n \to R^p \cup \{+\infty\}$ by

$$\tilde{f}(x) = \begin{cases} f(x) & \text{if } g(x) \leqq 0 \\ +\infty & \text{otherwise,} \end{cases}$$

the problem (P) can be rewritten as

$$\text{(P)} \qquad \text{minimize} \quad \tilde{f}(x).$$

A family of perturbed problems is provided by a function

$$\phi(x,u) = \begin{cases} f(x) & \text{if } g(x) \leqq u \\ +\infty & \text{otherwise,} \end{cases}$$

with the m-dimensional parameter vector u. Then the perturbation map W is

$$W(u) = \text{Min}\,\{\phi(x,u) \in R^p : x \in R^n\}$$

$$= \text{Min}\,\{f(x) : x \in R^n,\ g(x) \leqq u\}.$$

In this section we use the vector dual variable to define the dual problem and the Lagrangian, since we can obtain simpler expressions of them. For $\lambda \in R^m$,

$$-\phi^*(0,\lambda) = \text{Min}\,\{\phi(x,u) - \langle\langle\lambda,u\rangle\rangle \in R^p : x \in R^n,\ u \in R^m\}$$

$$= \text{Min}\,\{f(x) - \langle\langle\lambda,u\rangle\rangle : x \in R^n,\ u \in R^m,\ g(x) \leqq u\}.$$

Putting $u = g(x) + v$, we have

$$-\phi^*(0,\lambda) = \text{Min}\ \{f(x) - \langle\!\langle \lambda, g(x)\rangle\!\rangle - \langle\!\langle \lambda, v\rangle\!\rangle : x \in R^n,\ v \in R^m,\ v \geqq 0\}.$$

Hence, unless $\lambda \leqq 0$, $-\phi^*(0,\lambda) = \emptyset$. Moreover, for $\lambda \leqq 0$,

$$-\phi^*(0,\lambda) = \text{Min}\ \{f(x) - \langle\!\langle \lambda, g(x)\rangle\!\rangle : x \in R^n\}.$$

Thus, the dual problem with the vector dual variable can be written as

(D) $\qquad \text{Max} \bigcup_{\lambda \leqq 0} \text{Min}\ \{f(x) - \langle\!\langle \lambda, g(x)\rangle\!\rangle : x \in R^n\}.$

Theorem 1 provides the following duality theorem for multiobjective programming.

<u>Theorem 4</u>. (i) The problem (P) is stable if and only if, for each solution $\hat{x}$ to (P), there exists $\hat{\lambda} \in R^m$ with $\hat{\lambda} \leqq 0$ such that $\hat{\lambda}$ is a solution to the dual problem (D), $\langle \hat{\lambda}, g(\hat{x})\rangle = 0$, and

$$f(\hat{x}) \in \text{Min}\ \{f(x) - \langle\!\langle \hat{\lambda}, g(x)\rangle\!\rangle : x \in R^n\}.$$

(ii) Conversely, if $\hat{x} \in X$ and $\hat{\lambda} \in R^m$ with $\hat{\lambda} \leqq 0$ satisfy the above conditions, then $\hat{x}$ and $\hat{\lambda}$ are solutions to (P) and (D), respectively.

Proof: We prove that $\langle \hat{\lambda}, g(\hat{x})\rangle = 0$. Since $g(\hat{x}) \leqq 0$ and $\hat{\lambda} \leqq 0$, $\langle \hat{\lambda}, g(\hat{x})\rangle \geqq 0$. If $\langle \hat{\lambda}, g(\hat{x})\rangle > 0$, then $f(\hat{x}) > f(\hat{x}) - \langle\!\langle \hat{\lambda}, g(\hat{x})\rangle\!\rangle$, which contradicts the minimality of $f(\hat{x})$. Hence $\langle \hat{\lambda}, g(\hat{x})\rangle = 0$. The proof of remains is obvious.

In this case, the stability of the problem (P) is essentially guaranteed by convexity and Slater's constraint qualification.

<u>Proposition 16</u>. If each f_i $(i=1,\ldots,p)$ and g_j $(j=1,\ldots,m)$ is convex, if $W^*(\lambda)$ can be characterized completely by scalarization for each λ (for example, if every f_i is strictly convex), and if Slater's constraint qualification is satisfied (i.e., there exists $\bar{x} \in X$ such that

$g(\bar{x}) < 0$), then the problem (P) is stable.

Proof: We can directly apply Proposition 13.

Next we compute the Lagrangian with the vector variable.

$$L(x,\lambda) = \text{Min }\{\phi(x,u) - \langle\!\langle\lambda,u\rangle\!\rangle \in R^p : u \in R^m\}$$
$$= \text{Min }\{f(x) - \langle\!\langle\lambda,u\rangle\!\rangle : u \in R^m,\ g(x) \leqq u\}.$$

It is clear that $L(x,\lambda) = \emptyset$ for all x unless $\lambda \leqq 0$. On the other hand, when $\lambda \leqq 0$,

$$L(x,\lambda) = \{f(x) - \langle\!\langle\lambda,g(x)\rangle\!\rangle\} \quad \text{for all } x \in R^n.$$

Thus

$$-\phi^*(0,\lambda) = \text{Min} \bigcup_x L(x,\lambda).$$

In this case, a point $(\hat{x},\hat{\lambda})$ is a saddle point of L if and only if $\hat{\lambda} \leqq 0$ and

$$f(\hat{x}) - \langle\!\langle\hat{\lambda},g(\hat{x})\rangle\!\rangle \in [\text{Min} \bigcup_x L(x,\hat{\lambda})] \cap [\text{Max} \bigcup_\lambda L(\hat{x},\lambda)].$$

The relation

$$f(\hat{x}) - \langle\!\langle\hat{\lambda},g(\hat{x})\rangle\!\rangle \in \text{Max} \bigcup_\lambda L(\hat{x},\lambda)$$

is equivalent to the inequality

$$\langle\hat{\lambda},g(\hat{x})\rangle \leqq \langle\lambda,g(\hat{x})\rangle \quad \text{for all } \lambda \in R^m,\ \lambda \leqq 0.$$

Thus we have the following theorem.

Theorem 5. Suppose that the problem (P) is stable. Then $\hat{x} \in R^n$ is a solution to (P) if and only if there exists $\hat{\lambda} \in R^m$ such that $(\hat{x},\hat{\lambda})$ is a saddle point of the Lagrangian L, i.e.,

(i) $f(\hat{x}) - \langle\!\langle\hat{\lambda},g(\hat{x})\rangle\!\rangle \in \text{Min }\{f(x) - \langle\!\langle\hat{\lambda},g(x)\rangle\!\rangle : x \in R^n\}$;

(ii) $\hat{\lambda} \leqq 0$;

(iii) $\langle\hat{\lambda},g(\hat{x})\rangle \leqq \langle\lambda,g(\hat{x})\rangle$ for all $\lambda \in R^m$, $\lambda \leqq 0$.

7. Conclusion

In this paper we have overviewed some theoretical results with regard to conjugate duality in multiobjective optimization mainly according to Sawaragi et al.[11] and Tanino and Sawaragi[12]. The approach taken in this paper is based on efficiency (Pareto optimality) and some interesting results parallel with the well-known ordinary conjugate duality are obtained. However, there are some unsatisfactory points, since the concept of "vector supremum" is not well defined. Kawasaki[5,6] provided some interesting results by defining conjugates and subgradients via weak supremum. His approach is, however, artificial and so very difficult to understand. And weak Pareto optimality is not a so good solution concept as Pareto optimality. Hence we have adopted the more intuitive approach in this paper. The interested readers may refer to his papers or Sawaragi et al.[11]

References

1. Rockafellar, R.T., Convex Analysis, Princeton Univ. Press, Princeton, 1970.
2. Rockafellar, R.T., Conjugate Duality and Optimization, CBMS Lecture Notes Series No. 16, SIAM, Philadelphia, 1974.
3. Gros, C., Generalization of Fenchel's duality theory for convex vector optimization, Europ. J. Oper. Res., 2, 368, 1978.
4. Nieuwenhuis, J.W., Supremal points and generalized duality, Math. Oper. und Stat. Optimi., 11, 41, 1980.

5. Kawasaki, H., Conjugate relations and weak subdifferentials, Math. Opns. Res., 6, 593, 1981.

6. Kawasaki, H., A duality theorem in multiobjective nonlinear programming, Math. Opns. Res., 7, 95, 1982.

7. Ponstein, J., On the dualization of multiobjective optimization problems, Univ. of Groningen, Economic Institute Rep. 87, 1982.

8. Ponstein, J., Infima of sets, proper efficiency and multiobjective optimization, Univ. of Groningen, Economic Institute Rep. 88, 1982.

9. Zowe, J., A duality theory for a convex programming in order-complete vector lattices, J. Math. Anal. Appl., 50, 273, 1975.

10. Brumelle, S., Duality for multiobjective convex programs, Math. Opns. Res., 6, 159, 1981.

11. Sawaragi, Y., Nakayama, H. and Tanino, T., Theory of Multiobjective Optimization, Academic Press, New York, to appear.

12. Tanino, T. and Sawaragi, Y., Conjugate maps and duality in multiobjective optimization, J. Optimi. Theory Appl., 31, 473, 1980.

LINEAR MULTIPLE OBJECTIVE PROGRAMMING

Roger Hartley
Department of Decision Theory
University of Manchester

1. INTRODUCTION

Multiple objective optimisation has undergone considerable development in recent years and several approaches have been investigated. One of the closest to single objective optimisation is vector optimisation, in which efficient (non-dominated, admissible, Pareto optimal) solutions are sought. Since linear programming exhibits a particularly complete body of theory, we might expect the same to be true of vector linear programming and in this lecture we will describe some aspects of this theory.

To start, we will introduce some key definitions and terminology. There are three ingredients of a linear vector programming problem (LVP): a convex polyhedral set X in R^n (the intersection of a finite collection of half-spaces - not necessarily bounded), a linear mapping $C:R^n \to R^p$ and a convex cone $K \subset R^p$. The cone K induces an ordering on R^p, which we willwrite "$\geq_K$", by

$$x \geq_K y \quad \text{if and only if} \quad x-y\epsilon K.$$

The most frequently considered case is

$$K = \{x\epsilon R^p \mid x \geq 0\} \quad (=R^p_+)$$

and when making statements about this case we will suppress the reference to K. However, keeping K general permits us, for example, to impose restrictions on trade-offs between alternatives (giving a polyhedral K), to consider <u>weak efficiency</u> by taking

$$K = \{x\epsilon R^p \mid x_j > 0 \quad \text{for } j = 1, \ldots, p\} \quad (=R^p_{++}),$$

or to let K be the set of positive definite symmetric matrices of given size (of importance in statistics). The only imposition we will place on K is that it should have <u>strict support</u> in that the set K^q is non-empty, where K^q is the <u>strict polar core</u>:

$$K^q = \{x\epsilon R^p \mid x^T y > 0 \qquad \text{for all } y\epsilon K,\ y \neq 0\}.$$

Note that $(R^p_+)^q = R^p_{++}$ and $(R^p_{++})^q = R^p_+$.

The object, in vector programming, is to determine <u>efficient</u> points i.e. those $x\epsilon X$ for which any $y\epsilon X$ satisfying $Cy \geq_K Cx$ must also satisfy $Cx \geq_K Cy$. The <u>efficient set</u> is the set of all efficient points and is written E(X, C, K). It is worth noting that K induces an ordering on R^n defined by the convex cone.

$$\{x\epsilon R^n \mid Cx\epsilon K\} \quad (= K^*, \text{ say}).$$

It is clear that

$$E(X,C,K) = E(X,I,K^*)$$

where I is the identity mapping and it might seem that we could take C to be I without loosing generality or, equivalently, dispense with C altogether. However, whilst it is reasonable to assume K is strictly supported, this will typically not be valid for K^*.

Vector programming is closely related to parametric programming, as we shall see. A particular parametrisation has a single objective, formed by adding the weighted objective functions - the weights being the parameters. For any $w \in R^p$, we will define the single-objective problem LP(w) to be that of maximising $w^T Cx$ over $x \in X$. For any subset $S \subseteq R^n$, we will write M(X,C,S), for the set of maximisers in LP(w) for $w \in S$. That is

$$M(X,C,S) = \{x \in R^n | x \text{ is optimal in LP(w) for some } w \in S\}$$

In section 2, we will characterise E(X,C,K). Its relation to LP(w) is the key to this characterisation, and we will also investigate facial properties of E(X,C,K). In section 3, we will describe an algorithm for determining E(X,C,K) in a finite number of steps. Finally, section 4 looks at the stability of E(X,C,K), and its image under C in R^p, as a set-valued function of the data.

2. PROPERTIES OF E(X,C,K)

The relationship between efficient solutions of LVP and optimal solutions of LP(w) rests on the following separation theorem, a proof of which may be found in Hartley.[1]

THEOREM 2.1

Suppose that K_1 and K_2 are convex cones, with K_1 polyhedral and K_2 having strict support. If $K_1 \cap K_2 = \{0\}$, then there is a closed half-space H with $K_1 \subseteq H$ and $K_2 \cap H = \{0\}$.

Now suppose that $x \in E(X,C,K)$. This can be restated as a statement about sets in R^p, namely

$$CX \cap \{Cx + z \mid z \in K\} = \{Cx\} \qquad (2.1)$$

where

$$CX = \{Cy \mid y \in X\}.$$

Since X is polyhedral, so is CX (Stoer and Witzgall).[2] By translating the sets, we can assume that $Cx = 0$ without loss of generality. If Y is the convex cone generated by CX, we can translate (2.1) into

$$Y \cap K = \{0\}$$

We are now in a position to apply theorem 2.1 with $K_1 = Y$ (since Y is polyhedral) and $K_2 = K$ (assuming K is strictly supported). Suppose that the half-space H of the theorem is

$$H = \{z \in R^p \mid w^T z \le 0\}.$$

Then $Y \subseteq H$ says that $w^T z \leq 0$ for all $z \in Y$, which translates into x being optimal in LP(w), and $K \cap H = \{0\}$ says that $w \in K^q$.

Conversely, if x is optimal in LP(w) and $Cy \geq_K Cx$, where $w \in K^q$, then either $w^T(Cy - Cx) > 0$ or $Cy - Cx = 0$. However, the former contradicts the optimality of x, so the latter is true, putting $x \in E(X,C,K)$. These arguments support the following key result.

THEOREM 2.2

If K has strict support, then $E(X,C,K) = M(X,C,K^q)$.

Many consequences flow from this theorem and for the rest of this section we will assume K is strictly supported. For example, theorem 2.3 and its corollary follow directly.

THEOREM 2.3

E(X,C,K) is a union of faces of X.

COROLLARY 2.4

If F_1 and F_2 are efficient faces, then so is

(i) $F_1 \cap F_2$, and

(ii) F, where $F \subseteq F_1$

It follows from these results that a particularly economical description of E(X,C,K) is as the union of maximal efficient faces i.e. efficient faces properly contained in no other efficient face.

In the next section, we shall see how maximal efficient faces may be obtained algorithmically. For now, we look at another consequence of theorem 2.2 which has computational significance, as we shall see in Section 3.

A graph G(X, C, K) can be defined with the maximal efficient faces as vertices by including an edge $\{F_1, F_2\}$ if and only if F_1 and F_2 are distinct but not disjoint maximal efficient faces.

THEOREM 2.5

G(X,C,K) is connected.

We will indicate how this follows from theorem 2.2. Suppose F and G are maximal efficient faces and F is the optimal solution set of LP(w) and G of LP(v), where w, $v \in K^q$. For any $\lambda \in [0,1]$, we can define

$$w(\lambda) = \lambda v + (1-\lambda)w \in K^q$$

It is not hard to justify the following assertions.

(i) For any efficient face H, the set of λ for which H is a subset of the optimal solution set of LP(w(λ)) is a closed interval: I(H).

(ii) $0 \in I(F)$, $1 \in I(G)$

(iii) $I(H) \cap I(K) \neq \phi$ for efficient faces H and K implies $H \cap K \neq \phi$

These assertions can be seen to imply the existence of a sequence $F_1, \ldots, F_r$ of efficient faces with $F_1 = F$, $F_r = G$ and $F_s \cap F_{s+1} \neq \phi$, for $s = 1, \ldots, r-1$. The theorem follows by choosing a maximal efficient face containing F_s for each $s = 1, \ldots, r$.

The theorems above do not tell the whole story about E(X,C,K). For example, suppose C = I, p = 3 and

$$X = \{x \in R^3 \mid 2x_1 + x_3 \leq 1,\ -2x_1 + x_3 \leq 1,$$
$$2x_2 + x_3 \leq 1,\ -2x_2 + x_3 \leq 1,\ \ 0 \leq x_3 \leq 1\}$$

and consider the subset

$$\{x \in X \mid 2x_1 + x_3 = 1\} \ \cup\ \{-2x_1 + x_3 = 1\} \ \cup$$
$$\{2x_2 + x_3 = 1\} \ \cup\ \{-2x_2 + x_3 = 1\},$$

which satisfies the results of theorems 2.3 and 2.5. Nevertheless, it cannot be E(X,C,K) for any K, for, we would require

$$(2,0,1),(-2,0,1),(0,2,1),(0,-2,1) \in K^q$$

and therefore, by convexity, $(0,0,1) \in K^q$. But this would make the face $\{x \in X \mid x_3 = 1\}$ efficient. (We have assumed K to be strictly supported, but the conclusion holds without this assumption.) To the best of the author's knowledge no complete characterisation of E(X,C,K) for arbitrary K has been offered.

Having seen the importance of efficient faces, how can they be recognised? The answer to this question must depend on how a given face is described. To conclude this section, we will offer a characterisation of efficient faces described in terms of their vertices. (This is really only useful when X is bounded.) An alternative characterisation, more suited to computation, will be described in the next section.

LEMMA 2.6

Suppose $x^1, \ldots, x^r \in X$, $\lambda_s > 0$, $\mu_s \geq 0$ for $s = 1, \ldots, r$, and

$$\sum_{s=1}^{r} \lambda_s = \sum_{s=1}^{r} \mu_s = 1$$

Then

$$\sum_{s=1}^{r} \lambda_s x^s \in E(X,C,K) \text{ implies } \sum_{s=1}^{r} \mu_s x^s \in E(X,C,K)$$

Once again, this follows from theorem 2.2, for the hypothesis of the lemma implies that $\sum_{s=1}^{r} \lambda_s x^s$ is optimal in LP(w) for some $w \in K^q$. In particular, for $s = 1, \ldots, r$,

$$S \geq w^T C x^s$$

where

$$S = \sum_{s=1}^{r} \lambda_s w^T C x^s$$

Multiplying by λ_s and summing over $s = 1, \ldots, r-1$ gives

$$(1-\lambda_r)S \geq S - \lambda_r w^T C x^r,$$

so $w^T C x^r \geq S \geq w^T C x^r.$

By symmetry, $w^T C x^s = S$ for any s. Hence, for any $y \in X$,

$$\sum_{s=1}^{r} \mu_s w^T C x^s = S \geq w^T C y,$$

completing the proof, by theorem 2.2.

To use the lemma, we test the hypothesis by applying the Complementary Slackness Conditions [3] in LP(w). We will suppose that $a^1, \ldots, a^m \in R^n$ and

$$X = \{x \in R^n \mid (a^i)^T x \le b_i, \text{ for } i = 1, \ldots, m\}.$$

Then the i'th inequality is satisfied as an equality for $i \in I$, where

$$I = \{i = 1, \ldots, n \mid (a^i)^T x^s = b_i, \text{ for all } s = 1, \ldots, r\}.$$

The dual problem to LP(w) is

$$DP(w): \quad \text{minimise} \sum_{i=1}^{m} b_i y_i$$

$$\text{subject to} \sum_{i=1}^{m} a^i y_i = C^T w, \; y \ge 0.$$

This gives the following result.

THEOREM 2.7

For $x^1, \ldots, x^r \in X$, $H(x^1, \ldots, x^r) \subseteq E(X,C,K)$

if and only if there is a feasible solution of

$$\sum_{i \in I} a^i y_i = C^T w, \; y_i \ge 0 \text{ for } i \in I, \; w \in K^q,$$

where $H(x^1, \ldots, x^r)$ is the convex hull of $x^1, \ldots, x^r$.

3. AN ALGORITHM FOR CONSTRUCTING E(X,C,K)

In this section, we will sketch a method of obtaining the efficient set, roughly based on simplex ideas and utilising much of the groundwork prepared in Section 3. Some authors have to impose non-degeneracy restrictions on their algorithms. Validating such conditions computationally can be very difficult. These difficulties can be circumvented by dealing with index sets, a successful approach except, possibly, for highly degenerate problems.

We will suppose that X takes the form

$$X = \{x \in R^n \mid \sum_{j=1}^{n} A_j x_j = b, \quad x \geq 0\}$$

where $A_1, \ldots, A_n$, $b \in R^m$, which, of course, entails no real loss of generality. Any $J \subseteq \{1, \ldots, n\}$ defines a (possibly empty) face F_J, by

$$F_J = \{x \in X \mid x_j = 0 \quad \text{for } j \notin J\}$$

When $\{A_j\}$, for $j \in J$, is a basis of R^n, the set F_J is a singleton and therefore a vertex. We will call such a J a basic set and write x_J for the corresponding vertex. Of course, vertices and basic sets are determined by simplex tableaux.[3] Furthermore, every face (or vertex) is of the form F_J (or x_J) for some J, although, because of degeneracy, the correspondence is not necessarily 1-1. In many examples, some of the variables will be slack variables and the faces and vertices of interest will be those of the polyhedral set defined by the structural variables. This is equivalent to projecting the set X onto a subspace and it can be seen that subsets of $\{1, \ldots, n\}$ still determine faces and vertices, as above.

Suppose J is a basic set and let B be the matrix with columns A_j for $j \in J$ and C_B be the matrix with columns C_j for $j \in J$, where C_j is the j'th column of C. By analogy with simplex prices, we can associate a vector price Π_j with each non-basic variable x_j, for $j \notin J$, defined by

$$\Pi_j = C_B B^{-1} A_j - C_j$$

Note that $w^T \Pi_j$ is the price of x_j in LP(w).

We will say that the basic set J is efficient if there exists $w \in K^q$ making $w^T \Pi_j \geq 0$ for all $j \notin J$. Using the simplex method, lexigraphically modified to cope with degeneracy if necessary, [3] one can deduce the next result from theorem 2.2.

THEOREM 3.1

A vertex $x \in X$ is efficient if and only if there is an efficient vertex set J and $x = x_J$.

It follows from theorem 3.1 that finding all efficient basic sets also solves the problem of finding all efficient vertices.

A generalisation of this result can be used to characterise efficient faces. We will define $J \subseteq \{1, \ldots, n\}$ to be efficient if there is a $w \in K^q$ and a basic set $L \subseteq 1, \ldots, n$ with $L \subseteq J$ and $w^T \Pi_j \geq 0$ for $j \notin J$ and $w^T \Pi_j = 0$ for $j \in J \setminus L$. Note that L is efficient. The analogous result to theorem 3.1 is theorem 3.2.

THEOREM 3.2

A face F of X is efficient if and only if there is an efficient set $J \subseteq \{1, \ldots, n\}$ and $F = F_J$.

The method starts by constructing an efficient basic set. This can be done by finding an optimal solution of LP(w) for some $w \in K^q$. If X is bounded, an arbitrary $w \in K^q$ can be chosen but if X is unbounded it is possible for LP(w) to have no optimal solutions for certain $w \in K^q$, even if E(X,C,K) is non-empty. The boundedness of LP(w) can be assured by choosing $w \in K^q$ making the dual problem feasible. Using theorem 2.2, this gives the following result.

THEOREM 3.3

Suppose X is non-empty and consider the inequalities

$$(A_j)^T y \geq (C_j)^T w, \quad w \in K^q,\ j = 1, \ldots, n$$

(i) If $(\hat{y}, \hat{w})$ is a feasible solution of these inequalities then $LP(\hat{w})$ has an optimal (efficient) solution.

(ii) If the inequalities are infeasible, E(X,C,K) is empty.

This theorem can be used to find an efficient basic set by using the simplex method to solve LP($\hat{w}$) (or to show that the efficient set is empty if all LP(w) are unbounded). Given an efficient basic set L, the key step is to search for maximal efficient sets containing L, (these correspond to maximal efficient faces) which have not already been encountered. For any $j \in J \setminus L$, where J is a maximal efficient set, a new efficient basic set may be found by inserting x_j into the basis and using the usual pivot row selection rules to determine the existing variable. When no new efficient basic sets can be found in this manner, theorem 2.5 can be used to deduce that E(X,C,K) has been found - expressed as a union of maximal efficient faces.

This constitutes an extremely skeletal description of the algorithm and much ingenuity can be expended in avoiding duplication of effort and performing the calculations as effectively as possible. Some of the references cited below address this problem.

4. STABILITY

In this section, we investigate continuity properties of the efficient set and the set of efficient values. To this end, we now suppose

$$X \doteq \{x \in R^n \mid Ax \leq b,\ x \geq 0\}$$

and define two set-valued mappings. For any A,b,C, let

$\Phi(A,b,C) = E(X,C,R_+^p)$, and

$\Psi(A,b,C) = \{Cx | x \in \Phi(A,b,C)\}$

Note that we are assuming $K = R_+^P$. The results extend easily to polyhedral K and it is conjectured that they are also true for any strictly supported K.

We will take our definitions of continuity from Hogan.[4] They are widely used in mathematical programming. We will say that Φ is closed at $\bar{A}$, $\bar{b}$, $\bar{C}$ if, for any sequence A^s, b^s, C^s satisfying $(A^s, b^s, C^s) \to (\bar{A}, \bar{b}, \bar{C})$ as $s \to \infty$ and any sequence $x^s \in \Phi(A^s, b^s, C^s)$ satisfying $x^s \to \bar{x}$ as $s \to \infty$, we have $\bar{x} \in \Phi(A,b,C)$. We will say that Φ is open at $\bar{A}, \bar{b}, \bar{C}$ if, for any sequence A^s, b^s, C^s satisfying $(A^s, b^s, C^s) \to (\bar{A}, \bar{b}, \bar{C})$ and any $\bar{x} \in \Phi(\bar{A}, \bar{b}, \bar{C})$, there is an integer $M \geq 1$ and a sequence $x^s \in \Phi(A^s, b^s, C^s)$ for $s \geq M$, satisfying $x^s \to \bar{x}$ as $s \to \infty$. When Φ is both open and closed, we will say that it is continuous. Similar definitions apply to Ψ.

In order to obtain general results, we need to apply conditions to LVP. We will say that the problem is P-regular at $\bar{A}, \bar{b}, \bar{C}$, if the system

$$\bar{A}x \leq 0, \quad \bar{C}x = 0, \quad x \geq 0,$$

has no non-zero solution and that the problem is D-regular, if the system

$\bar{A}^T y \geq 0, \quad \bar{b}^T y = 0, \quad y \geq 0$

has no non-zero solution. A problem which is both P- and D-regular is called <u>regular</u>

All of the following theorems are proved in Hartley and Rustichini.[5] Unfortunately, the proofs are rather involved and too lengthy to reproduce here. We start by giving the most general result.

THEOREM 4.1

If LVP is regular at $\bar{A},\bar{b},\bar{C}$, then Φ is open at $\bar{A},\bar{b},\bar{C}$.

In testing for regularity, it is useful to note that the existence of $\bar{x}$ satisfying $A\bar{x} \leq 0$, $\bar{x} \geq 0$ and $\bar{x} \neq 0$ implies that the feasible region is unbounded. Furthermore, suppose $\hat{x} \geq 0$ satisfies $(\bar{a}^i)^T\hat{x} < b_i$ for $i = 1, \ldots, m$, where $\bar{a}^i$ is the i'th row of $\bar{A}$ and that $\bar{y} \geq 0$ satisfies $\bar{A}^T\bar{y} \geq 0$ and $\bar{y} \neq 0$, then

$$\bar{b}^T\bar{y} > \bar{y}^T\bar{A}\hat{x} \geq 0.$$

These observations justify the following conditions.

THEOREM 4.2

(i) Boundedness of the feasible region implies P-regularity

(ii) An $\hat{x} \geq 0$, satisfying $(\bar{a}^i)^T\hat{x} < b_i$ for all i implies D-regularity.

From linear programming theory, we might expect Ψ to be continuous. The next example shows that this expectation cannot be fulfilled.

EXAMPLE 4.3

$$\bar{A} = (1,1), \quad \bar{b} = 1, \quad \bar{C} = \begin{vmatrix} 1 & 1 \\ 1 & -1 \end{vmatrix}$$

Choosing, for $s \geq 1$,

$$A^s = (1 + s^{-1}, 1), \; b^s = \bar{b}, \; C^s = \bar{C}$$

it is readily verified from theorem 2.2, that $(0,1)\in\Phi(A^s,b^s,C^s)$ and thus $(1,-1)\in\Psi(A^s,b^s,C^s)$ for $s \geq 1$. However, $\Psi(1,-1)\in(\bar{A},\bar{b},\bar{C})$. even though regularity can be confirmed via theorem 4.2.

Even in the single objective case, D-regularity is necessary to ensure continuity of the objective function (Martin,[6] Bereanu[7]). Example 4.4 shows that when $p \geq 2$, P-regularity is essential for Ψ to be open.

EXAMPLE 4.4

$$\bar{A} = (0,1), \; \bar{b} = 1, \; \bar{C} = \begin{pmatrix} 0 & -1 \\ 0 & 2 \end{pmatrix}$$

Note that the problem is D-regular by theorem 4.2. Observe that $(0,0)\in\Psi(\bar{A},\bar{b},\bar{C})$, so $(0,0)\in\Phi(\bar{A},\bar{b},\bar{C})$. Choosing

$$A^s = \bar{A}, \quad b^s = \bar{b}, \quad C^s = \begin{pmatrix} s^{-1} & -1 \\ -s^{-1} & 2 \end{pmatrix}, \text{for } s \geq 1,$$

it can be shown that $(z_1, z_2) \epsilon \Phi(A^s, b^s, C^s)$ implies $z_1 + z_2 = 1$. Hence Ψ is not open.

We now turn to the case when A is fixed and we are interested in continuity of $\Phi(b,C)$ and $\Psi(b,C)$. Example 4.4 shows that regularity of some sort is required for openness. Indeed we have the next result.

THEOREM 4.5

If A is fixed and LVP is P-regular at $\bar{A},\bar{b},\bar{C}$, then Ψ is open at $A,\bar{b},\bar{C}$.

Regular examples can be constructed, in which Ψ is not closed and Φ is neither open nor closed. This is in striking contrast to single objective problems, where Φ is closed and the objective function value is continuous, even without any regularity assumptions.

The one case in which the results for vector problems exactly parallel those for single objective problems is when A and C are fixed.

THEOREM 4.6

If A and C are fixed, Φ and Ψ are continuous in b.

5. BIBLIOGRAPHY

The preceding development, particularly in Sections 2 and 3, derives ideas from the work of many authors and in this section we will give a short and selective bibliography.

A proof of the fundamental Theorem 2.2 when $K = R^P_+$ can be found in Philip [6] along with several other useful results. Another characterisation of efficient faces may be found in Bitran and Magnanti [7] and characterisations of maximal efficient faces appear in Ecker et al. [8] and Isermann [9], Evans and Stoer [10] and Ecker and Kouada[11] describe multiple objective versions of the simplex method. We have assumed that the only task of such a method is to generate the efficient set, but if it is also considered desirable to identify the set of weights for which any given efficient face is optimal, methods based on multiparametric analysis such as that of Yu and Zeleny [12] may be more appropriate.

Many papers on the problem of finding an initial efficient vertex have been published, not all of them correct. The procedure implicit in the use of Theorem 3.3 involves the solution of two linear programming problems when the feasible region is not known to be bounded. Ecker and Kouada [13] exhibit a single problem which has an optimal solution (which is efficient) if and only if the efficient set is non-empty. The solution will typically not be a vertex, so further computation is required to obtain a basic set. [14]

6. REFERENCES

1. Hartley, R. On cone-efficiency, cone-convexity, and cone compactness, *SIAM.J.Appl.Maths.* 34 (1978), 211-222.

2. Stoer, J. and Witzgall, C. *Convexity and Optimisation in Finite Dimensions*, Springer, Berlin 1970.

3. Hartley, R. *Linear and Nonlinear Programming*, Ellis Horwood, Chichester, England. to appear (1985)

4. Hogan, W. W. Point-to-set-maps in mathematical programming, *SIAM Review*, 15 (1973) 591-603.

5. Hartley, R. and Rùstichini, A. *Stability in Multicriteria Linear Programming*, Preprint, University of Manchester.

6. Philip, J. Algorithms for the vector maximum problem, *Math.Prog.*, 2 (1979) 207-229.

7. Bitran, G. R. and Magnanti, T. L., *Duality based Characterisation of Efficient Facets*, MIT Prepring (1979).

8. Ecker, J. E. Hegner, N. S. and Kouada, I.A. Generating all maximal efficient faces for multiple objective linear programs, *J.O.T.A.*, 30 (1980), 353-381

9. Isermann, H. The Enumeration of the set of all efficient solutions for a linear multiple objective program, *O.R.Q.*, 28 (1977), 711-725.

10. Evans, J. P. and Steuer, R. E. A revised simplex method for linear multiple objective programs, *Math.Prog.* 5, (1973) 54-72.

11. Ecker, J. E. and Kouada, I.A. Finding all efficient extreme points for multiple objective linear programs, *Math.Prog.* 14 (1978) 249-261.

12. Yu, P. L. and Zeleny, M. The set of all non-dominated solutions in linear cases and a multi-criteria simplex method, *J.Math.And.Appl.* 49 (1975), 430-468.

13. Ecker, J. E. and Kouada, I.A. Finding efficient points for linear multiple objective programs, *Math.Prog.* 8, (1975) 375-377.

14. Ecker, J. E. and Jegner, Nancy S., On computing an initial efficient extreme point, *J.O.R.S.* 31 (1980), 591-594.

SOME EXISTENCE RESULTS AND STABILITY IN MULTI OBJECTIVE OPTIMIZATION

Roberto Lucchetti
Istituto di Matematica
University of Genoa

If someone is curious about multiobjective optimization, and enters in a library to look for works in the argument, he can find an enormous amount of things, results, applications, references. For this and other reasons, it is quite obvious that the results given here are a very little part of the known theorems, mainly in the existence. So, no claims to be complete. We shall deal with a topological vector space X and a cone $C \subset \mathbb{R}^m$ that is always assumed closed convex, pointed ($C \cap -C = \{0\}$) and with nonempty interior. Moreover it is given a function $f : X \longrightarrow \mathbb{R}^m$ to be maximized with respect to the ordering induced by the cone, on a general constraint set A. We are interested in existence theorems, namely the non-emptiness of the set $S_C = \{ x \in A : [f(x) + C] \cap f(A) = f(x)\}$. At first, observe that, if $C_1 \subset C_2$, then $S_{C_1} \supset S_{C_2}$: this means that, in particular, the non emptiness of S_C guarantees existence of solutions for the weak optimization, namely of the points $x \in A$: $f(x) + \overset{o}{C} \cap f(A) = \emptyset$. In the sequel we write S for S_C and W for the weak solutions. In the scalar case, i.e. when m = 1 the most celebrated theorem says that, if f is upper semicontinuous (u.s.c.) and there is in A a relatively compact maximizing sequence, then there is a solution for the problem. We want to present here an analogous simple result, so we have to clarify the meaning of "maximizing sequences" and "upper semicontinuity". We shall make use of the following notations:

C^o is the polar cone of C, $C^o = \{p \in \mathbb{R}^m : (p, c) \geq 0 \quad \forall c \in C\}$.
G(C) is a family of unitary of generators of C, and $y <_C x$ means that $x - y \in C$. C(x) is the set of the elements comparable with x : $C(x) = \{y \in \mathbb{R}^m : y - x \in C \cup (-C)$. Observe that $x >_C 0$ if and only if $(x, p) \geq 0 \ \forall p \in G(C^o)$: so $G(C^o)$ has the meaning of the "cohordinate" versors related to C. In particular, if $C = \mathbb{R}^m_+$ $x \,_C\!> 0$ iff $x_i \geq 0$ for every i = 1 .... m.

At first, we want to introduce the concept of sup of a subset of $\mathbb{R}^m$, and as in the case m = 1 we add to $\mathbb{R}^m$ two elements, called $-\infty$ and $+\infty$, with the following natural properties:

$\forall a > 0, \quad \forall b < 0, \quad \forall p \in G(C^o), \quad \forall x \in \mathbb{R}^m$

$a(\pm\infty) = \pm\infty$, $b(\pm\infty) = \mp\infty$, $0(\pm\infty)$ 0, $+\infty - \infty = -\infty$, $x \pm\infty = \pm\infty$, $(p, \pm\infty) = \pm\infty$, $-\infty <_C x <_C +\infty$.

Call $\bar{\mathbb{R}}^m$ this set: to have a topology on it we consider:

$S(x_o, s) = \{x \in \mathbb{R}^m : |x_i - x_{oi}| < s_i\}$

$S^+(x_o) = \{x_o - \mathscr{C} C \cup +\infty\}$

$S^-(x_o) = \{x_o + \mathscr{C} C \cup -\infty\}$

As $x_o \in Q^m$, $s \in Q^m_+$ it can be shown that we have a basis for a topology τ in $\bar{\mathbb{R}}^m$ in such a way that $(\bar{\mathbb{R}}^m, \tau)$ is compact and separable, satisfies the axioms of countability, but it is not Hausdorff. However there is only one case in which a converging sequence x_n has more than one limit: precisely when $\lim x_n = \{+\infty, -\infty\}$ (example: $x_n = (-n, n)$ with $C = \mathbb{R}^2_+$). In such a case if will be useful to put:

LIM x_n = lim x_n if it is unique

LIM $x_n = -\infty$ if not.

We are now able to give the following definitions:

$A \subset \overline{\mathbb{R}}^m$, $\sup A = \{ y : y >_C x \quad \forall x \in A \cap C(y)$ and there is $x_n \in A \cap C(y) : \text{LIM } x_n = y \}$; $\max A = A \cap \sup A$.

Observe that there are other possible and suitable definitions of sup, mainly in duality theory (See the work of Tanino in this book). For extrema problems for instance there is the following definition due to Cesari, Suryanarayana[1]: $\sup' A = \{ x \in \bar{A} : \nexists\, y \in A : y >_C x \text{ and } y \neq x \}$.

It is obvious that $\sup' A \supset \sup A$, but $\max' A = \max A$: this means that if an element of $\sup' A$ is in A, it is also in sup A. Moreover the choice of the LIM to define the sup is motivated by the fact that, having to "maximize", we do not want that "too many" sets have $+\infty$ in the sup. (See example below).

We have the following, expected, proposition: $\forall A \neq \emptyset \quad \sup A \neq \emptyset$ and $\forall x \in A \quad \exists \; y \in \sup A : y >_C x$. Now, a maximizing sequence can be defined as a sequence converging (in the sense of the LIM) to an element of the sup.

Upper semicontinuity (in sequential sense, sometimes).

There is not only one way to extend the definition of u.s.c. for a vector valued function: for instance f is (seq.) u.s.c. at a point x_o if

a) $\forall x_n \longrightarrow x_o$: $\lim f(x_n) \neq \emptyset$ then $\text{LIM } f(x_n) <_C f(x_o)$.

b) $\forall x_n \longrightarrow x_o$, $\quad \forall \varepsilon > 0$, $\quad \forall h \in \text{int } C$ (or $\exists \; h \in \text{int } C$) s.t. $|h| = 1$ there is k : $\forall n > k \quad f(x_n) <_C f(x_o) + \varepsilon h$ if $f(x_o) > -\infty$ or $f(x_n) <_C - \varepsilon h$ if $f(x_o) = -\infty$.

c) all the components of f are u.s.c. at x_o. Namely $(f(\cdot), p)$ is u.s.c. at $x_o \quad \forall p \in G(C^\circ)$.

Moreover, u.s.c. at every point can be stated

d) $\{ (x, y) \in X \times \mathbb{R}^m : f(x) >_C y \}$ is closed

e) $\{ x \in X : f(x) >_C M \}$ is closed $\quad \forall M \in \mathbb{R}^m$.

In the case m = 1 and if x_o is arbitrary, it is well known that a).....e) are all equivalent. This is not longer true if m > 1, and the situation can vary if the cone C is finitely generated or not. However a), at every x_o, is always equivalent to d) and e) and it is the weakest between those considered.

It is the time to state the theorem:

Let $f : X \longrightarrow \bar{\mathbb{R}}^m$ such that:

$f(x) \neq +\infty \quad \forall x \in X$ and there is $x_o : f(x_o) \neq -\infty$.

f satisfies a at every point (u.s.c.)

$$K = \bigcap_{\substack{p \in C^o \\ |p| = 1}} \left\{ x \in X : (f(x), p) \geqslant m \right\} \text{ is compact } \forall m \in \mathbb{R}$$

(rel. compactness of maximizing sequences).

Then $\forall y \in \sup \left\{ f(x) : x \in X \right\}$ there is $x \in X : f(x) = y$.

Observe that the constraint set A can be ignored as f can assume the value $-\infty$, and that if A is compact, the condition on K is automatically fulfilled.

Consider the following simple example: $X = \mathbb{R}$ $m = 2$ $C = \mathbb{R}^2_+$ $f(x) = (x,-x)$. Then $S = \mathbb{R}$, but the theorem could be false if we allow that $+\infty$ is in the sup of $f(\mathbb{R})$. There is a sequence in $f(\mathbb{R})$ converging to $\left\{ +\infty , -\infty \right\}$ and this motivates the choice of LIM in the definition of the sup.

For other results about existence, see for instance Corley[2], where the statement is the same for what concerns u.s.c. but it is given a slightly different notion of compactness. Observe however that here it is not supposed int $C \neq \emptyset$. See also the references cited there. We want to mention also Borwein[3], where the results are given in terms of properties related directly to the image set, and there is a deep study on the relationships between existence, properties of the cone and the choice axiom. The result presented here can be found in Caligaris, Oliva[4], while Caligaris, Oliva[5] study the case $C = \mathbb{R}^m_+$ and Caligaris, Oliva[6] the situation when $\mathbb{R}^m$ is substituted by an infinite dimensional reflexive separable Banach space. Caligaris, Oliva[7] analize the relationships between several possible extension of u.s.c. to the vectorial case: this is made in part by Trudzik[8] where convexity and semicontinuity are also related to the continuity.

Stability

Suppose to have $f : X \times P \longrightarrow \mathbb{R}^m$, $K : P \rightrightarrows X$ (multifunction). P is the parameter space and for some $p_o \in P$ we can consider the initial maximum problem presented before. Write $S(p_o)$ and $W(p_o)$ for the strong and weak solutions, and $f(S(p_o), p_o)$, $f(W(p_o), p_o)$ for the related values.

As usual for stability we mean to find conditions on f and K guaranteeing "stable" or "good" behavior of the optimal sets and values. For the scalar case, a lot of results are well-known: some of them can be found in Lucchetti[9]. For the vectorial case we have the following results: (for therminology about multifunctions, see Lucchetti[10]). For W: if f is continuous, K lower semicontinuous (l.s.c.) and closed, then W is closed (nothing can be stated about optimal values!). If f is continuous, K l.s.c., u.s.c. and compact valued, then W, and $f(W, \cdot)$ are u.s.c.

Observe that these results are direct extensions of the scalar case.

About S simple examples show that it is not possible to find results on the solutions with a rather general function f : so the properties are stated in terms of values, or supposing, if you want, $X = \mathbb{R}^m$ and $f(x, p) = x \;\forall p$.

In any case the results are in apposite sense; namely:

K compact valued, l. and u.s.c., or K convex valued, l.s.c. and closed, then S is l.s.c.

For all these results, see Lucchetti[10] where it is also presented an application of the previous results to an economical context, perhaps one of the first fields of research in which multiobjective optimization (here called also Pareto maximization) was used. It is shown that natural hypotheses on the convergence of the data defining the economies give closedness of the set of the efficient allocations of the market. Moreover if we add the suitable hypotheses guaranteeing existence of an equilibrium in a free disposal economy with production, we get the much stronger property of u.s.c. In such a way it is also possible to get u.s.c. of the equilibria of the market merely showing the property of closedness. This is shown in Lucchetti, Patrone[11].

To conclude, we recall some other results about stability:

Tanino, Sawaragi[12], where it is possible to find results in the same direction of those here presented, but in a more general setting: namely to dominance relation is not given, for every point, by the fixed cone, but it can vary in a general way: however the authors need further compactness hypotheses in order to get closedness or l.s.c. results that in Lucchetti[10] are avoided.

Jansen, Tijs[13] and Jurkiewicz[14] investigate continuity properties of particular points belonging to the Pareto boundary of a set: this study is motivated by the fact that these point are meaningful for instance for bargaining problems in game theory.

Peirone[15] starts the study of Γ convergences for vectorial valued functions: the reason is that, in the scalar case, Γ convergence is suitable (and rather general) to get stability results.

Finally in the article of this book of prof. Hartley other results about stability are shown, mainly in linear programming.

I want to thank prof. Jahn and Tanino, and dr. Tardella of the University of Pisa, for some helpful discussions.

REFERENCES

1 Cesari, L. - Suryanarayana, M.B., Existence Theorems for Pareto Optimization in Banach Spaces, Bull. Amer. Math. Soc. (82), 1976.

2 Corley, H.V., An Existence Result for Maximizations with Respect to Cones, J. Optim. Th. and Appl. (31) 1980.

3 Borwein, J., On the Existence of Pareto Efficient Points, Math. of Oper. Res. (8), 1983.

4 Caligaris, O. - Oliva, P., Necessary and Sufficient Conditions for Pareto Problems, Boll. U.M.I. (17 B), 1980.

5 Caligaris, O. - Oliva, P., Optimality in Pareto Problems, Proceedings in "Generalized Lagrangians in Systems and Economic Theory", IIASA Luxembourg, 1979.

6 Caligaris, O. - Oliva, P., Constrained Optimization of Infinite Dimensional Vector Functions with Application to Infinite Horizon Integrals, to appear.

7 Caligaris, O. - Oliva, P., Semicontinuità di Funzioni a Valori Vettoriali ed Esistenza del Minimo per Problemi di Pareto, Boll. U.M.I. (II C) 1983.

8 Trudzik, L.I., Continuity Properties for Vector Valued Convex Functions, J. Austral. Math. Soc. (36) 1984.

9 Lucchetti, R., On the Continuity of the Value and of the Optimal Set in Minimum Problems, Ist. Mat. Appl. C.N.R. Genova, 1983.

10 Lucchetti, R., Stability in Pareto Problems, to appear.

11 Lucchetti, R. - Patrone, F., Closure and Upper Semicontinuity Results in Mathematical Programming, Nash and Economic equilibria, to appear.

12 Tanino, M.J. - Sawaragi, Y., Stability of Nondominated Solutions in Multicriteria Decision Making, J. Optim. Th. and Appl. (30), 1980.

13 Jansen, M.J. - Tijs, S., Continuity of the Barganing Solutions, Intern. J. of Game Th. (12), 1983.

14 Jurkiewicz, E., Stability of Compromise Solution in Multicriteria Decision Making Problems, J. Optim. Th. and Appl. (31), 1980.

15 Peirone, R., Γ limiti e minimi Pareto, Atti Accad. Naz. Lincei (LXXIV), 1983.

APPLICABILITY OF THE FUNCTIONAL EQUATION IN MULTI CRITERIA DYNAMIC PROGRAMMING

Mordechai I. Henig
Faculty of Management
Tel Aviv University

INTRODUCTION

Dynamic programming (DP) is associated with mathematical techniques to optimize decision problems by recursively solving a functional equation. When a problem is sequential over time, such an equation arises naturally. In other cases, stages are introduced to faciltate such an equation, although the stages have no natural order. An example of the latter case is the allocation of a resource to the production of several items, where each stage is associated with an item.

The domain of the functional equation is a state space considered to be the sufficient information at each stage to make an optimal decision. To validate that this equation solves the problem, the principle of optimality, as stated by Bellman[1] is invoked: "An optimal policy has the property that whatever the initial node (state) and initial arc (decision) are, the remaining arcs (decisions) must constitute an optimal policy with regard to the node (state) resulting from the first transition". This principle which appears to be a "natural truth" distinguishes DP not only as a technical tool but also as an approach toward optimization. Indeed, DP is both similar and different from other models and techniques in operations research. Like other models in mathematical programming it has an objective function and possibly explicit constraints as well. On the other hand, it is characterized by decisions sets which reflect implicit constraints. Furthermore, outcomes may be stochastic as well as deterministic. Comprehensive discussions and references can be found in recent books on DP by Denardo[2] and by Heyman and Sobel[3].

Introducing multicriteria into this framework may require special techniques which are not common in multicriteria mathematical programming. To avoid discussing the fundamental question of how to make decisions over time, but rather to concentrate on the problem arising from the multiplicity of criteria, we assume that by considering each criterion separately the functional equation is a possible presentation of the problem, and that it can be solved efficiently. The question on which we focus in this article is: To what extent can we generalize this equation when several criteria are judged together, and how can it be solved efficiently? More specifically the following issues are directly or indirectly considered:

- How and when do multicriteria problems arise in the context of DP?
- The implication of assuming the von Neumann utility function on solving the functional equation, and on the articulation of

preferences by the decision maker (DM).

- The efficiency of interactive procedures to find optimal decisions with relation to the degree of accuracy required.

The article is divided into two sections. In the first, a stochastic model is discussed. Its application lie in decision problems over time, where at each time period some uncertainty is resolved and a decision has to be made.

The topic of utility functions over time, especially when streams of income and consumption are considered, has attracted a lot of attention in the literature. For instance, Chapter 9 of Keeney and Raiffa[4]. Multiple criteria decision making over time is a relatively new area of research mainly because of its stochastic elements. Among the recent books on multiobjective methods, we found only "some thoughts on future developments" (Chapter 12 of Goicoechea et al.[5]). Papers, mainly in Markov decision processes, have been published by Furukawa[6], Hartley[7], Henig[8], Viswanathan et al.[9], Shin[10] and White and Kim[11].

It is well known that the application of the functional equation is restricted to "separable" utility functions. Practically speaking, utility function over time is either additive or multiplicative. Furthermore, due to the stochastic elements, the local (one-stage) utility functions are linear. If these conditions are acceptable, a search for an optimal policy becomes relatively simple. Otherwise no general method of solution exists.

The second section deals with multicriteria network models. Research in this subject has mainly been conducted under the titles of ratio optimization or optimization under constraints. Most of the papers addressed the problem of efficiency, notably Megido[12], Chandrasekaran[13] and Handler and Zang[14]. Hansen[15] directly analyzed the multicriteria problem in networks and found that by extending the algorithms their polynomial efficiency may be lost. One criterion DP and network theory meet in what is called the shortest path problem, hence multi criteria shortest path problem is the focus of this section. Papers on this subject were published by Hartley[16], Daellenbach and Kluyrer[17], Climaco and Martins[18], Henig[19] and Warburton[20]. It is shown that non-separable utility functions can be optimized by applying the functional equation. The major obstacle, however, may be that most existing algorithms are not polynomially bounded. Some ideas on how efficiency can be obtained are presented.

A STOCHASTIC MODEL

The multicriteria DP model consists of:

S - set of <u>states</u>;

$D = \{D(s): s \in S\}$ - a collection of <u>decision</u> sets;

$W = \{w(s,d): s \in S, d \in D\}$ - a collection of sample spaces (each in a probability space);

t: $S \times D \times W \to S$ - a <u>transition</u> function; and

$\tilde{r} = (\tilde{r}^1,\ldots,\tilde{r}^m)$: $S \times D \times W \to R^m$ - an <u>immediate returns</u> function, i.e., $\tilde{r}(s,d,\cdot)$ is an m-dimensional random variable.

The diagram below depicts the rules of the process.

For simplicity of notation and discussion we assumed a finite stationary process. A <u>policy</u> is a rule according to which a decision is selected - given all the information about the process. A policy is

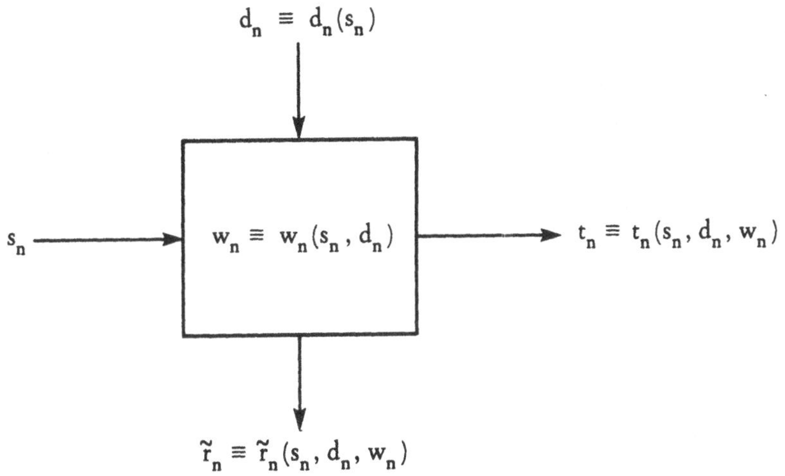

called memoryless if the state and stage of the process are the sufficient information to make an optimal decision. If the policy involves a random decision selection it is called randomized otherwise it is pure.

A standard example is an inventory problem where $s \in S$ is the level of inventory. A decision $d \geq 0$ is the amount ordered, w is the stochastic demand and $t(s,d) = s + d - w$. In the one criterion model the immediate return is the profit $\tilde{r}^1(s,d,w) = pw - h(s+d)$ where p is the unit of profit and h the inventory unit cost. Another attribute related with inventory management is the possibility of shortage, which cannot be fully measured by an immediate drop in profit. Thus the DM may consider another immediate return $\tilde{r}^2(s,d,w) = \min\{0, s + d - w\}$.

Returning to the general model, the DM at stage 1, confronts a stochastic, multiperiod and multicriteria problem (technically, a period can be considered as a criterion but we avoid such an approach). The DM has to select a decision for every possible state (or at least for s_1) such that multicriteria random stream of outcomes will be optimal (or at least satisfying).

A reasonable study before making a decision is to obtain some information from the DM concerning preferences. This can be done only if some minimal set of axioms or preconditions are assumed. Here, we assume the (implicit) existence of the von-Neumann-Morgenstern notions and concepts of utility function, and that the DM wants to maximize its expected value. Mathematically speaking, the DM wants to find the policy which solves

$$\bar{U}_1(s_1) = \max_D \{ \mathop{E}_{w_1, w_2, \ldots, w_N} [U(\tilde{r}_1, \tilde{r}_2, \ldots, \tilde{r}_N)] \} . \quad (1)$$

The existence of such a function U and the use of expectation follow some very reasonable assumptions (Fishburn[21]). However, the difficulty in this approach is the assessment of U (Keeney and Raiffa[4]). One of the roles of multicriteria methods is to overcome these difficulties. For example, interaction methods enable the DM to articulate his preferences by confronting possible outcomes of his decisions. We emphasize here the interaction approach although other multicriteria methods may be applied as well. Several interactive techniques are reported in the literature and for details the reader is referred to Goicoechea et al.[5], Chankong and Haimes[22], Zionts[23] and White[24].

According to interaction procedures the DM is presented with possible outcomes which serve to extract more information about his preferences which in turn are used to obtain other outcomes. This is repeated until satisfactory outcomes are obtained. But what outcomes can be presented in our dynamic stochastic problem? By applying the principle of optimality we can decompose the N periods problem to N one-period problems, thereby presenting the DM with a one period outcome (successively for N periods) rather than a vector of N outcomes over time. A known result in DP is that the principle of optimality is generally satisfied only when the utility function over time is separable, which in most applications means an additive or a multiplicative function. The reason for this restriction is that in other types of utility, history-remembering policies should be considered as well (Kreps[25]). To use a functional equation which can be solved by some tractable method we must assume that (positive) local utility functions $U_n(\tilde{r}_n)$, n=1,2,...,N exist such that

$$U(\tilde{r}_1,\tilde{r}_2,\ldots,\tilde{r}_N) = \sum_{n=1}^{N} U_n(\tilde{r}_n) \quad , \quad \text{or} \tag{2}$$

$$U(\tilde{r}_1,\tilde{r}_2,\ldots,\tilde{r}_N) = \prod_{n=1}^{N} U_n(\tilde{r}_n) \ . \tag{3}$$

The functional equation for every $s \in S$ is

$$\overline{U}_n(s) = \max_{d \in D(s)} \{E_{w_n} [U_n(\tilde{r}_n(s,d,w_n)) + \overline{U}_{n+1}(t(s,d,w_n))]\}, \tag{4}$$

$$n=1,\ldots,N-1$$

$$\overline{U}_N(s) = \max_{d \in D(s)} \{E_{w_N} [U_N(\tilde{r}_N(s,d,w_N))]\}$$

where + is the additive or multiplicative operator.

If the local utility functions can be assessed then (4) may be solved by standard mathematical optimization techniques. Otherwise multicriteria methods may be used, and again the question is what outcomes per stage can be presented to the DM. Since expected utility is to be maximized, we would like to reveal the DM's preferences about expected outcomes. However, generally it is not valid that

$$E_{w_n}[U_n(\tilde{r}_n(s,d,w_n))] = U_n(r(s,d))$$

where we denote $r(s,d) = E_{w_n} \tilde{r}(s,d,w_n)$. The only case when such a relation is valid (since the criteria are dependent within one stage) is for a linear function, i.e.,

$$U_n(\tilde{r}_n(s,d,w)) = \sum_{\ell=1}^{m} \beta_n^{\ell} \tilde{r}_n^{\ell}(s,d,w) = \beta_n \tilde{r}_n(s,d,w)$$

The vector β_n of <u>weights</u> has the interpretation of the relative importance of the criteria. Under this assumption (4) becomes

$$\bar{U}_n(s) = \max_{d \in D(s)} \{\beta_n r_n(s,d) + E \bar{U}_{n+1}(t(s,d,w_n))\} \tag{5}$$

$$\bar{U}_N(s) = \max_{d \in D(s)} \{\beta_N r_N(s,d)\}$$

If $\bar{U}_{n+1}(s)$ is known for every $s \in S$ and β_n is given then solving $\bar{U}_n(s)$

is a standard one-stage optimization. If β_n is unknown then by interaction with DM, several β_n vectors are generated, an optimal decision is found for each, until DM converges to the preferred one. At each iteration of this procedure the DM is presented with $r_n(s,d)$ and $\underset{w_n}{E}\,\overline{U}_{n+1}(t(s,d,w_n))$ and (possibility) the weights β_n.

When multiplicative utility is assumed then

$$\overline{U}_n(s) = \max_{d\epsilon D(s)} \{ \sum_{\ell=1}^{m} \beta_n^\ell [r_n^\ell(s,d) \cdot E\,\overline{U}_{n+1}(t(s,d,w_n))]\} \ , \ n=1,\ldots,N-1$$

and when m=2

$$\{r_n^1(s,d) \cdot E\,\overline{U}_{n+1}(t(s,d,w_n)),\ r_n^2(s,d) \cdot E\,\overline{U}_{n+1}(t(s,d,w_n)) : d\epsilon D(s)\}$$

for a given $s \in S$ can be depicted in R^2. Similarly, if $\alpha_n = \sum_{\ell=1}^{m} \beta_n^\ell$ is known in the additive case then

$$\overline{U}_n(s) = \max_{d\epsilon D(s)} \{ \sum_{\ell=1}^{m} \beta_n^\ell [r_n^\ell(s,d) + \frac{1}{\alpha_n} E\,\overline{U}_{n+1}(t(s,d,w_n))]\} \ ,$$

$$n=1,\ldots,N-1$$

and when m=2 we can depict

$$\{r_n^1(s,d) + \frac{1}{\alpha_n} E\,\overline{U}_{n+1}(t(s,d,w_n)),\ r_n^2(s,d) + \frac{1}{\alpha_n} E\,\overline{U}_{n+1}(t(s,d,w_n)) : d\epsilon D(s)\}$$

for a given $s \in S$ in R^2.

The vulnerable spot in this procedure is that $\bar{U}_{n+1}$ has to be known explicitly in stage n. In other words β_N, $\beta_{N-1},\ldots,\beta_{n+1}$ have to be assessed before $\bar{U}_n(s)$ can be solved. This difficulty can be circumvented if the weights can be separated into two factors α_n - the weight of stage n and $\beta^1,\ldots,\beta^m$ the stationary, relative importance of the criteria, so that $\beta_n^\ell = \alpha_n \cdot \beta^\ell$. Usually $\alpha_n = \rho^n$ where ρ is the discount factor. If α_n is known then (5) is a standard function equation for each β and $U_n(s)$ can be solved recursively to obtain an optimal policy as a function of β. Thus for each β an optimal vector of objective functions is obtained. Each of them is an expected (discounted) total of the immediate returns. By varying β and by presenting the optimal vector of objective functions (for the state s_1) in R^m, the DM can converge to the preferred policy.

In summary, the main observation is that the assumptions about separability and linearity simplified a complicated problem involving multicriteria multi-period under uncertainty to a one period deterministic multi-criteria problem.

What if the utility function is not separable? Since this question concerns one-criterion problems, the answer can be found in the standard DP literature (Hinderer[26] and Kreps[25]). Generally speaking the optimal policy may not be memoryless or pure. In other words, the sufficient information to select an optimal decision is not merely the state and the stage of the process, but also its past history. Conceptually, nothing is wrong, however, the functional equation must be solved for every possible history of the process and this is usually intractable.

Can we replace expectation with other random utility measures?

Again, the answer can be found in the literature of one-criterion DP. For most practical problems, expectation is the only operator which is workable with the functional equation (because it is a linear operator), although some secondary measures are possible (Jaquette[27]). For example, Sobel[28] considered a linear combination of expectation and variance as a measure but showed that the functional equation is not valid (see also Miller[29]).

In these cases when additivity, linearity and expectation criterion are not satisfied, solving the functional equation may yield only an approximate optimal decision whose accuracy cannot be estimated.

The question of solving a one criterion functional equation under "dynamic" constraints has attracted some researchers' attention. Rossman[30] and Derman[31] formulated such models and concluded that an optimal policy may be randomized. As indicated by Sneidovich[32] and Beja[33], this is not acceptable. In our terms Rossman and Derman utility functions were defined over the objective functions (in R^2), where each of them is the expected (discounted) total of immediate returns. In both models immediate returns are "benefit" and "reliability" (the first is expressed as a functional equation and the second as a "dynamic" constraint). However, unless the utility function is separable, the use of the functional equation is not justified, thereby leading both authors to recommend a randomized policy. As demonstrated by Beja, a chance-constrained formulation does not faithfully reflect the DM's utility function.

It is worth noting that a dynamic constraint may play an important role in communicating with the DM who finds it difficult to express his utilities in terms of weights. Thus interaction with the DM can be expressed in terms of level of the constraint, but the optimal policy can be located among the memoryless pure policies by varying the vector of β until the DM is satisfied with the level of constraint which such a policy achieves.

A DETERMINISTIC MODEL

In a deterministic model the transition as well as the local return are soley functions of the state and action. For simplicity we assume finite sets of stages, states and decisions.

An example is the knapsack problem where $\sum_{n=1}^{N} p_n x_n$ is to be maximized over

$$\{x_n : \sum c_n x_n \le b,\ x_n \text{ and } c_n \text{ are positive integers}\}$$

The state space is $\{(n,y):n=1,\ldots,N;\ y=0,1,\ldots,b\}$, $d(n,y) = \{x=0,1,2\ldots : c_n x \le y\}$, $t((n,y),x) = (n-1,y-c_n x)$ and $r_n^1(n,y,x) = p_n x$. At stage 1 the state is $s_1 = (N,b)$. A policy indicates an amount x for every state (n,y), and is associated with a feasible solution of the problem. The coefficients p_n and c_n are the profit and volume per unit of item n. In a bicriterion case $r_n^2((n,y),x) = r_n x$, where r_n may be the weight per unit of item n, and the problem is both to maximize total profit and to minimize total weight.

Another example is the well-known shortest path problem where a finite set N of nodes (states) and a set, $A \subset N \times N$, of directed arcs (decisions), are given. Two nodes, denoted by O and D, are specified, indicating the origin and the destination of the network. Each arc $(i,j) \in A$ is associated with a vector $r(i,j) = (r^1(i,j),\ldots,r^m(i,j))$ of immediate returns (e.g., cost, time, hazard). A path p (from i_0 to i_k) is a sequence of arcs $\{(i_0,i_1),\ldots,(i_{k-1},i_k)\}$. Let P_i be the set of paths from i to D and let $P = P_0$ be the set of policies (paths from 0 to D). DM wants to maximize the utility function

$$u(p) = u[(r^1(0,i_1),\ldots,r^m(0,i_1)),\ldots,(r^1(i_k,D),\ldots,r^m(i_k,D))] \qquad (6)$$

(where $p = ((0,i_1),\ldots,(i_k,D)) \in P$) over all $p \in P$.

Denote $\bar{u}_0 = \max_{p\in P} u(p)$. In order to use a functional equation to find $\bar{u}_0$ consider two different cases. In the first assume that local utility functions u_i, $i \in S$, exist such that

$$\bar{u}_i = \max_{j\in S}\{u_i(r^1(i,j),\ldots,r^m(i,j)) + \bar{u}_j\}\ ,\ i \in S\ ,\ \bar{u}_D = 0 \qquad (7)$$

This functional equation is similar to (4) in the stochastic model. Since no randomization is involved then the DM can be presented with vectors $r(i,j) = (r^1(i,j),\ldots,r^m(i,j))$ to select the optimal node j. However, as in the stochastic model the values of $\bar{u}_j$ have to be known in order to calculate $\bar{u}_i$. On the other hand, u_i can be a nonlinear function.

When u_i is explicit, then solving (7) for every $i \in S$ is the usual shortest path algorithm except that a calculation of u_i is required. If u_i is revealed during an interaction procedure then solving (7) depends on how the DM selects the preferred decision, and how u_i is revealed. In a case of pairwise comparisons between decisions (i.e., arcs emanating from i) the complexity of the shortest path algorithm is retained since this is exactly the routine of selecting the minimum value applied in the shortest path algorithm.

So far we have assumed the existence of local utility functions which depend on the state (node) of the process. In other words, the vector of immediate returns in the various arcs along the path were grouped according to nodes. Another possibility is to group them according to objective functions, i.e.,

$$u(p) = u(f^1(p),\dots,f^m(p)) \tag{8}$$

where $f^\ell(p) = \sum_{(i,j)\epsilon p} r^\ell(i,j)$ (multiplication can be considered by taking log of r^ℓ). The domain of u is R^m and in the general case maximizing u(p) over $p \in P$ is difficult. If u is monotone (decreasing) in every component then the optimal path has objective functions values which are nondominated. To generate the set of nondominated paths is possible (see Daellenbach and Kluyrer[17]), but it requires a huge amount of time and storage. Furthermore, the number of nondominated paths may be large and a search for the optimal path has to be conducted.

If u has other properties in addition to being monotone, they can be applied to accomodate certain interactive methods so that calculations and storage can be reduced considerably.

One way to exploit such properties is to consider the network problem as a multiobjective mathematical programming with linear constraints. A standard result in network theory is the modelling of the shortest path problem as a linear programming problem. Thus by weighting the various objective functions we get a linear programming model whose basic solutions are equivalent to the paths. Any path which uniquely minimizes a weighted objective function is called <u>extreme</u> since its objective function values are an extreme point of the convex hull of $\{f^1(p);,\ldots,f^m(p): p \in P\}$.

When u is quasiconvex a standard result in nonlinear programming is an optimal path which is also extreme. By varying the weights, generating extreme paths and presenting them to the DM we may converge to the optimal path. We illustrate such an interactive procedure for m=2. The procedure is adapted from a branch and bound algorithm for explicit quasiconvex u suggested by Henig[19].

Iteration K ($K \geq 2$) of the procedure, begins with a list of (nondominated) extreme paths $p_k: k=1,\ldots,K\}$. Each path is associated with the values (f_k^1, f_k^2) of the objective functions, such that $f_1^1 > f_2^1 > \ldots > F_K^1$, and with a list b_k, $1 > b_k > 0$, such that

$$a_k = b_k f_k^1 + (1-b_k) f_k^2 < b_k f^1(p) + (1-b_k) f^2(p) \quad \forall \quad p \in P.$$

Let (g_k^1, g_n^2), $k=1,\ldots,k-1$ be the solution of

$$b_k g_k^1 + (1-b_k) g_k^2 = a_k$$
$$b_{k+1} g_n^1 + (1-b_{k+1}) g_k^2 = a_{k+1} \; .$$

Clearly the hyperplane $H = \bigcap_{k=1}^{K} \{(g^1,g^2): b_k g^1 + (1-b_k) g^2 \geq a_k\}$ contains the convex hull of $\{f^1(p),\ldots,f^m(p)\colon p \in P\}$. Furthermore, $\{(g_k^1,g_k^2)\colon k=1,\ldots,K-1\}$ are the extreme points of H and therefore u attains its maximum there. The DM is presented with

$$\{(f_k^1,f_k^2)\colon k=1,\ldots,K\} \quad \text{and} \quad \{(g_k^1,g_k^2)\colon k=1,\ldots,k-1\}.$$

If some (f_k^1,f_k^2) is most preferred, then p_k is the optimal path. If some (g_k^1,g_k^2) is most preferred then (g_k^1,g_k^2) is removed and for

$$b = (f_{k+1}^2 - f_k^2)/(f_k^1 - f_{k+1}^1 + f_{k+1}^2 - f_k^2)$$

$a = \min_{p \in P}\{bf^1(p) + (1-b)f^2(p)\}$ is found.

If $a = bf_k^1(1-b)f_k^2$ then no new point is added. If $bf_k^1 + (1-b)f_k^2 > a = bf^1(p^*) + (1-b)f^1(p^*)$ for some $p^* \in P$ then $(f^1(p^*), f^2(p^*))$ is the K+1 extreme path and next iteration starts.

In the worst possible case all the extreme paths are generated before the DM is able to select the preferred path, however, usually many (g^1,g^2) type of points are not "branched" (i.e., they are not selected) because they are "bounded" by (i.e., they are inferior to) extreme paths.

When u is quasiconcave the optimal path is not necessarily an extreme one. However, if restricted to extreme paths only then relatively efficient interactive procedures can be employed to recover the best extreme path. For example, the Zoints-Wallenius[34] method or the general search method of Geoffrion et al.'s[35] can be used. For m=2 a binary search as suggested by Henig[19] can be used. However, further search among non-extreme paths is needed if the optimal path is wanted. That leads us to a different way of how properties of u can be used to reduce complexity.

The theme of this second way is the application of methods unique to integer programming and specially to network problems to the multicriteria case. For example, to find the optimal path when u is quasiconcave the algorithm to find the k-th (k=1,2,...) shortest path can be used. The idea is that if the optimal extreme path was found then its objective function values are not far away from those of the optimal path.

Let p_1 be the best extreme path with values $(f_1^1,\ldots,f_1^m)$ such that $bf_1 \leq bf(p)$ for every $p \in P$. Then by weighting each arc by the vector $(b^1,b^2,\ldots,b^m)$ the k-th shortest path k=2,3... can be found successively (Lawler[36]). Notice that paths whose values are dominated can be generated, and clearly, they are not presented to the DM. This algorithm is polynomially efficient if k is known, however, in our case this number is a priori unknown. Experiments with explicit utility functions showed that k was relatively small (Handler and Zang[14]).

An important feature of many algorithms related to network structure is their polynomial complexity. Such algorithms to find the shortest path were developed (except for the linear case), only for the case of ratio (of linear functions) optimization. Some other cases were proven to be NP-complex problems, among them the constrained shortest path. Thus for the general utility function, even under convexity conditions, it is reasonable that a polynomially bounded algorithm does not exist. In interactive procedures it means that the number of calculations, time spent with the DM and storage requirements may be too large even if the network is of moderate size.

A method which does allow us to control the complexity of the algorithm is available if the DM is willing to compromise some degree of accuracy. Let p^* be the optimal path. A path p is δ-optimal if $f(p) \leq f(p^*)(1+\delta)$ for $0 < \delta < 1$.

The method suggested by Warburton[20] actually uses the functional equation to find the set of non-dominated paths after modifying the immediate returns as functions of δ. The larger δ the smaller the set of nondominated paths obtained. The major thrust of the method is that the relationship between δ and the number of calculations is known a priori. The ability to reduce complexity stems from the fact that when immediate returns are integers, the number of nondominated paths is limited by

$$\alpha = \min[f^1(p_1) - \max_{\ell \neq 1} f^1(p_\ell),\ldots,f^m(p_m) - \max_{\ell \neq m} f^m(p_\ell)]$$

where p_ℓ is the path which minimizes the ℓ-th objective function. Thus by an appropriate scalarization of the immediate returns as a function of α and δ and rounding off, the number of nondominated paths can be controlled and δ-optimality is achieved.

We shall conclude with an application of Megido's[12] polynomial polynomial algorithm to an interactive procedure which finds an optimal path when m=2 under the following conditions:

(i) u is linear, i.e.,

$$u(f^1(p),f^2(p)) = b^*f^1(p) + (1-b^*)f^2(p)$$

for some $0 < b^* < 1$; however, b^* is unknown.

(ii) Given $p_1, p_2 \in P$ such that

$$p_i \text{ minimizes } b_i f^1(p) + (1-b_i)f^2(p), \quad i=1,2, \text{ over } p \in P$$

with $b_1 > b_2$, then $b^* > b_2$ if the DM prefers p_1 over p_2 and $b^* < b_1$ otherwise.

The algorithm starts with node D, $\underline{b} = 0$, $\overline{b} = 1$ and then recursively goes over all nodes while narrowing the interval $[\underline{b},\overline{b}]$. Let $\overline{N} \subset N$ and suppose that $\underline{b}$ and $\overline{b}$ $(0 \le \underline{b} \le \overline{b} \le 1)$ are given such that for every $j \in \overline{N}$ there exists a path $p_j \in P_j$ whose value of objective functions $(f^1(p_j),f^2(p_j))$ are the (unique) values of the shortest path from j to D with respect to every $b \in [\underline{b},\overline{b}]$. Let $G_i(b) = bf^1(p_i) + (1-b)f^2(p_i)$ be a function over $[0,1]$ and

$$\overline{G}_i(b) = \min\{br^1(i,j) + (1-b)r^1(i,j) + G_j(b) : j \in \overline{N}\} .$$

Clearly, $\overline{G}_i(b)$ is a concave piecewise linear function over $[\underline{b},\overline{b}]$. If it is linear it means that $j^* \in \overline{N}$ exists such that $\overline{G}_i(b) = br^1(i,j^*) + (1-b)r^2(i,j^*) + G_{j^*}(b_i)$ for every $b \in [\underline{b},\overline{b}]$. In this case let $G_i(b) = \overline{G}_i(b)$, $p_i = \{(i,j^*) \cup p_{j^*}\}$, $\overline{N} = N \cup \{i\}$ and continue with the next node in $N - \overline{N}$.

If $\overline{G}_i(b)$ is not linear over $[\underline{b},\overline{b}]$ then there exists nodes j and k, paths $p_j \in P_j$, $p_k \in P_k$ and $\tilde{b} \in [\underline{b},\overline{b}]$ such that

$$\overline{G}_i(\tilde{b}) = \tilde{b}r^1(i,j)+(1-\tilde{b})r^2(i,j)+G_j(\tilde{b}) = \tilde{b}r^1(i,k)+(1-\tilde{b})r^2(i,k)+G_k(\tilde{b}) .$$

Denote $p^1 = (i,j) \cup p_j$ and $p^2 = (i,k) \cup p_k$, and find the shortest path q from 0 to i with respect to the weight b (i.e., each arc (i,j) is associated with $\tilde{b}r^1(i,j) - (1-\tilde{b})r^2(i,j)$). Let $q_1 = q \cup p_1$, $q_2 = q \cup q_2$ and $(f^1(q_1),f^2(q_1))$ and $(f^1(q_2),f^2(q_2))$ be their respective objective function values. Suppose that $f^1(q_1) < f^1(q_2)$. If the DM prefers q_1 then $\underline{b} = \tilde{b}$, and $p_i = p^1$, otherwise $\overline{b} = \tilde{b}$ and $p_i = p^2$. The algorithm is continued with a narrower $[\underline{b},\overline{b}]$.

The algorithm terminates after $0 \in \overline{N}$. The number of calculations in an acyclic network with N nodes is bounded by $O(n^4)$, and the DM has to make no more than $n^2/2$ comparisons.

This algorithm with minor modifications can be used to find the best extreme path when u is quasiconcave.

CONCLUSIONS

In this article we discussed the restrictions imposed on the application of the functional equation to solve multicriteria decision problems.

In the stochastic case linear local utility functions were assumed in order to maximize expectation of a separable utility function over time. Under these conditions, which are common in practice, it is clear that the DM will be satisfied with a memoryless and pure policy. Even if the DM expresses his preferences via non-separable utility, he may still be satisfied with such a policy if the outcomes are favorable.

When a non-separable utility or another measure than expectation is assumed, a revision in the model, usually in the state space, is necessary before calculations can begin. When the local utility functions are nonlinear, or in a case of nonstationary weights, the search for an optimal policy is more complicated, and further theoretical research is required to assess these utilities via multicriteria methods.

In the deterministic case a variety of utility functions can be considered. However, solving the functional equation and searching for the optimal policy may be expensive. We showed how properties of the utility function and of the network structure can be used to reduce time and search cost. Clearly, there is much ground for further research. To mention a few possibilities: to generalize the algorithms mentioned for more than two criteria, to develop other methods of approximation and to find algorithms whose expected complexity (as opposed to the worst case analyzed so far) is polynomial.

References

1. Bellman R., Dynamic Programming, Princeton University Press, Princeton, New Jersey, 1957.
2. Denardo E.V., Dynamic Programming, Theory and Applications, Prentice Hall, 1982.
3. Heyman D. and M. Sobel, Stochastic Models in Operations Research, Vol. II, McGraw-Hill, 1984.
4. Keeney R. and H. Raiffa, Decisions With Multiple Objectives: Preferences and Value Tradeoffs, John Wiley & Sons, 1976.
5. Goicoechea A., D. Hansen and L. Duchstein, Multiple Decision Analysis With Engineering and Business Applications, John Wiley & Sons, 1982.
6. Furukawa N. , Characterization of optimal policies in vector valued Markovian decision processes, Mathematics of Operations Research 5 (2), 271, 1980.
7. Hartley R., Finite, discounted vector Markov decisions processes, Notes in Decision Theory, Note 85, University of Manchester, 1979.
8. Henig M., Vector-valued dynamic programming, SIAM Journal of Control and Optimization, 420, 1983.
9. Viswanathan B., V.V. Aggarwal and K.P.K. Nair, Multiple criteria Markov decision processes, TIMS Studies in the Management Sciences 6, North-Holland Publishing Company, 263, 1977.
10. Shin M., Computational methods for Markov decision problems, Ph.D. Dissertation, University of British Columbia, 1980.

11. White C. and K. Kim, Solution procedures for vector criterion Markov decision processes, Large Scale Systems 129, 1980.

12. Megido N., Combinatorial optimization with rational objective function, Mathematics of Operations Research, 4, 414, 1979.

13. Chandrasekaran R., Minimal ratio spanning trees, Networks, 335, 1977.

14. Handler G. and I. Zang, A dual algorithm for the constrained shortest path problem, Networks 10, 293, 1980.

15. Hansen P., Bicriterion path problems, in Multiple Criteria Decision Making Theory and Application Proceedings, 1979 by Fandel G. and Gal T., eds., Saringer Verlag, 1980, 109.

16. Hartley R., Dynamic programming in vector networks, Notes in Decision Theory, Note 86, University of Manchester, 1979.

17. Daellenbach H.G. and D.C.D. De Kluyrer, Note on multiple objective dynamic programming, Journal of Operational Research Society 31, 591, 1980.

18. Climaco J. and E. Martins, A bicriterion shortest path algorithm. European Journal of Operations Research, 11, 399, 1982.

19. Henig M., The shortest path problem with two objective functions, Working Paper 743/82, Tel Aviv University, 1982.

20. Warburton A., Exact and approximate solution of multiple objective shortest path problems. Working Paper 83-74, University of Ottowa, 1983.

21. Fishburn P.C., Utility Theory for Decision Making, John Wiley & Sons, 1970.

22. Chankong V. and Y. Haimes, Multiobjective Decision Making Theory

and Methodology, North-Holland, 1983.

23. Zoints S., Multiple criteria decision making: An overview and several approaches, Working Paper 454, State University of New York at Buffalo, 1980.

24. White D.J., The foundation of multi-objective interactive programming - Some questions, Notes in Decision Theory, Note 126, 1982.

25. Kreps D., Decision problems with expected utility criteria I: Upper and lower convergent utility, Mathematics of Operations Research, 2, 45, 1977.

26. Hinderer K., Foundations of Non-Stationary Dynamic Programming with Discrete Time Parameter, Springer-Verlag, Berlin, 1970.

27. Jaquette S., Markov decision processes with a new optimality criterion: Discrete time, The Annals of Statistics, 1(3), 496, 1977.

28. Sobel M., The variance of discounted Markov decision processes. Journal of Applied Probability 9, 794, 1983.

29. Miller B.L., Letter to the editor. Management Science, 24, 1979, 1978.

30. Rossman L., Reliability-constrained dynamic programming and randomized release rules in reservoir management, Water Resources Research, 13 (2), 247, 1970.

31. Derman C., Optimal replacement and maintenance under Markovian deterioration with probability bounds on failure, Management Science, 9(3), 478, 1963.

32. Sneidovich M., Chance constrained dynamic programming,

RC 7459 IBM Thomas J. Watson Research Center, Yorktown Heights, N.Y. 1978.

33. Beja A., Probability bounds in replacement policies for Markov systems, Management Science, 16(3), 253, 1969.

34. Zionts S. and J. Wallenius, On interactive methods for solving the multiple criteria problem, Management Science, 22(6), 652, 1976.

35. Geoffrion A.M., J.S. Dyer and A. Feinberg, An interactive approach for multicriterion optimization with an application to the operation of an academic department, Management Science 19(4), 357, 1972.

36. Lawler E., Combinatorial Optimization, Holt, Rinehart and Winston, 1976.

VECTOR OPTIMAL ROUTING BY DYNAMIC PROGRAMMING

Roger Hartley
Department of Decision Theory
University of Manchester

1. Introduction

Algorithms for generating shortest paths have been widely studied in the combinational optimisation literature for many years. More recently, the vector version, in which non-dominated paths are sought, has been investigated. Acyclic problems were discussed by Randolph and Ringeisen[1] and Thuente.[2] The bicriterion case was analysed by Hansen[3] and extended by Climaco and Martins[4] and Martins.[5] A multicriterion algorithm was given by Martins.[6] White[7] used linear programming methodology but was forced to exclude non-dominated paths which were dominated by a convex combination of path lengths.

None of these authors uses dynamic programming although this was one of the earliest techniques applied to the sealar case (Bellman[8]). It is still computationally valuable when some lengths are negative - as may occur when the algorithm is used as a subroutine in another procedure. In this paper we offer a vector generalisation of dynamic programming for optimal routing. The scheme is rather similar to one of White[9] for which, however, no theory is developed.

2. Notation, terminology and statement of the problem

A vector network consists of a directed graph with finite vertex sets S and, for each arc (i,j), a p-component vector $\underset{\sim}{c}(i,j)$. Let t be a vertex in S which we call the *target* vertex. We are interested in paths in the network terminating at t. With path P we can associate a vector length $\underset{\sim}{c}(P)$ by adding $\underset{\sim}{c}(i,j)$ over the arcs of P. For each $i \in S$ $(i \neq t)$ we will write (i) for the set of paths from i to t and we will call $P \in \mathcal{P}(i)$ *efficient* if there is no other $Q \in \mathcal{P}(i)$ satisfying $\underset{\sim}{c}(Q) < \underset{\sim}{c}(P)$ where this notation is equivalent to $\underset{\sim}{c}(Q) \leq \underset{\sim}{c}(P)$ (componentwise ordering) and $\underset{\sim}{c}(Q) \neq \underset{\sim}{c}(P)$. We denote this set of efficient paths by E(i).

We now introduce some notation which will be used throughout the paper together with some conventional assumptions about the directed graph underlying the network. The first such assumption is that for every $i \in S$, $(i \neq t)$ there is at least one path from i to t. This assumption guarantees, *inter alia*, that E(i) is non-empty for all $i \in S$ $(i \neq t)$. The second is that arcs of the form (i,i) are excluded.

For any i we denote by $\Gamma(i) \in S$ the set of j such that (i,j) is an arc and so our second assumption can be rephrased as $i \notin \Gamma(i)$. Our third and final assumption is that $\Gamma(t)$ is empty. If this assumption is not satisfied *a priori* we can always add an extra vertex t' and a single arc (t,t') with $\underset{\sim}{c}(t,t') = 0$. We then take t' as target vertex. Clearly this change does not affect efficient paths, other than by adding the arc (t,t') to all such paths. If our network has the property that for any circuit K we have $\underset{\sim}{c}(K) \geq 0$ we can drop the third assumption. This applies *a forteriori* if $\underset{\sim}{c}(i,j) \geq 0$ for all arcs.

For any set A of p-vectors we write eff A for the set of efficient members of A i.e.

$$\text{eff } A = \{\underset{\sim}{a} \in A \mid \text{there is no } \underset{\sim}{b} \in A \text{ with } \underset{\sim}{b} < \underset{\sim}{a}\}.$$

If $\underset{\sim}{x}$ is a p-vector $\underset{\sim}{x}+A$ denotes the set

$$\underset{\sim}{x}+A = \{\underset{\sim}{y} \mid \underset{\sim}{y} = \underset{\sim}{x}+\underset{\sim}{a} \text{ for some } a \in A\}$$

Finally, we will assume S has N members.

3. A computational scheme

We will consider the following scheme in which the sets $V_n(i)$ for $i \in S$ and $n = 1,2,\ldots.$ are defined iteratively as follows ($\underset{\sim}{0}$ is a p-vector).

$$V_o(i) = \{\infty\} \text{ for all } i \in S, \ (i \neq t)$$

$$V_n(t) = \{\underset{\sim}{0}\} \text{ for } n = 0,1,2,\ldots.$$

$$V_n(i) = \text{eff } \cup\{\underset{\sim}{c}(i,j) + V_{n-1}(j) \mid j \in \Gamma(i)\} \text{ for } i \in S \ (i \neq t) \text{ and } n = 1,2,\ldots.$$

where ∞ is a conventional element satisfying $\infty > \underset{\sim}{x}$ and $\infty+\underset{\sim}{x} = \infty$ for all vectors $\underset{\sim}{x}$. This convention avoids over-complicating the notation. For any $i \in S$, $(i \neq t)$ we write $\mathcal{P}_n(i)$ for the set of paths from i to t containing n or fewer arcs and $E_n(i)$ for the set of efficient paths with respect to $\mathcal{P}_n(i)$ i.e. $P \in E_n(i)$ if and only if $P \in \mathcal{P}_n(i)$ and there is no $Q \in \mathcal{P}_n(i)$ with $\underset{\sim}{c}(Q) < \underset{\sim}{c}(P)$.

Our first result states that the recessive procedure correctly obtains $E_n(i)$.

Theorem 1 For all $i \in S$ $(i \neq t)$ and $n = 1,2,\ldots.$

$$E_n(i) = \{P \in \mathcal{P}_n(i) \mid \underset{\sim}{c}(P) \in V_n(i)\}$$

Vector dynamic programming was considered by Yu and Seiford[10] who showed that the principle of optimality need not always hold and by Hartley[11] who gave a counterexample even when the combination operation between stages is addition (see also White[12]). However, the method works here because we are dealing with finite sets, the crucial point being that any inefficient path is dominated by an efficient path. The simple proof of this assertion and of theorem 1 is left to the reader.

Our next result shows that, provided all circuits have non-negative length, the computational scheme is finite, where a *circuit* is a path from a vertex to itself.

Theorem 2 The following statements are equivalent.

(a) $\underset{\sim}{c}(K) \geq 0$ for every circuit K (componentwise inequality).

(b) $V_N(i) = V_{N-1}(i)$ for all $i \in S$.

Theorem 2 is proved using lemma 1, whose proof is left to the reader.

Lemma 1 Suppose A and B are finite sets of p-vectors and for any $\underset{\sim}{a} \in A$ there is a $\underset{\sim}{b} \in B$ satisfying $\underset{\sim}{b} \leq \underset{\sim}{a}$, whilst for any $\underset{\sim}{b} \in B$ there is $\underset{\sim}{a} \in A$ satisfying $\underset{\sim}{a} \leq \underset{\sim}{b}$. Then eff A = eff B.

Proof of theorem 2 (a) => (b)

Choose $i \in S$ $(i \neq t)$ and consider the following sets of p-vectors

$$A = \{\underset{\sim}{c}(P) \mid P \in \mathcal{P}_{N-1}(i)\}$$

$$B = \{\underset{\sim}{c}(P) \mid P \in \mathcal{P}_{N}(i)\}.$$

Since $\mathcal{P}_{N-1}(i) \subseteq \mathcal{P}_N(i)$, it follows at once that for any $\underset{\sim}{a} \in A$ there is $\underset{\sim}{b} \in B$ with $\underset{\sim}{b} \le \underset{\sim}{a}$. Conversely, suppose $\underset{\sim}{b} = \underset{\sim}{c}(P) \in B$ where $P \in \mathcal{P}_N(i)$. Then, either $P \in \mathcal{P}_{N-1}(i)$ in which case we put $\underset{\sim}{a} = \underset{\sim}{b}$ or P has N+1 vertices. In the latter case some vertex must be repeated in P which means P contains a circuit. Removing this circuit leaves a path $Q \in \mathcal{P}_{N-1}(i)$ and by (a) in the theorem we have $\underset{\sim}{c}(Q) \le \underset{\sim}{c}(P)$. Put $\underset{\sim}{a} = \underset{\sim}{c}(Q)$. Whichever way $\underset{\sim}{a}$ is defined we have $\underset{\sim}{a} \in A$ and $\underset{\sim}{a} \in A$ and $\underset{\sim}{a} \le \underset{\sim}{b}$. Hence eff B = eff A. The result follows via theorem 1.

<u>(b) => (a)</u>

We suppose (b) holds but not (a) and derive a contradiction. Suppose K is a circuit and $c_r(K) < 0$ where c_r is the r'th component of $\underset{\sim}{c}$. Choose a vertex i of K ($i \neq t$). By the assumptions described in the previous section, we can construct, for any $m \ge 1$, a path from i to t which consists of traversing m times round K and then following any path from i to t. Therefore we can find a $P \in \mathcal{P}_n(i)$ for some n such that $c_r(P) < v_r$ for all $\underset{\sim}{v} \in V_{N-1}(i)$. Hence $V_n(i) \neq V_{N-1}(i)$, contradicting $V_N(i) = V_{N-1}(i)$.

We will now examine how E(i) may be constructed. The proof is ommitted.

<u>Theorem 3</u> Assume (a) in theorem 2 is valid. Then, for any $i \in S$ ($i \neq t$)

$$E(i) = \{P \in \mathcal{P}(i) \mid \underset{\sim}{c}(P) \in V_{N-1}(i)\}$$

Provided condition (a) of theorem 2 is valid, for any $P \in E(i)$ there is $Q \in E_{N-1}(i)$ with $\underset{\sim}{c}(Q) = \underset{\sim}{c}(P)$ and so we first give a method for generating $E_{N-1}(i)$ for $i \in S$ ($i \neq t$). We note that all $P \in E(i)$ which do not contain a circuit are in $E_{N-1}(i)$.

Routine A

1. Put $i_1 = i$ and $r = 1$. Choose $\underset{\sim}{v} \in V_{N-1}(i)$.

2. Choose $j \in \Gamma(i,j)$, $\underset{\sim}{w} \in V_{N-r-1}(j)$ so that $\underset{\sim}{v} = \underset{\sim}{c}(i_r,j) + \underset{\sim}{w}$.

3. Change r to r+1. Put $i_r = j$ and $\underset{\sim}{v} = \underset{\sim}{w}$. If $i_r = t$, go to 4.

   Otherwise, go to 2.

4. $P = (i_1,\ldots,i_r)$ is efficient.

Theorem 4A

Routine A, with all possible choices in steps 1 and 2 generates $E_{N-1}(i)$.

To find E(i) we must use *Routine B* which is identical to routine A except that in step 2 we replace $V_{N-r-1}(j)$ by $V_{N-1}(j)$.

Theorem 4B

Routine B, with all possible choices in steps 1 and 2 which cause the routine to terminate, generates E(i).

We note that if, for some $n < N$ we find $V_n(i) = V_{n-1}(i)$ for all $i \in S$, then we must have $V_N(i) = V_{N-1}(i) = V_n(i)$ and we may terminate our calculations at this point.

4. A Counterexample

The algorithm described in section 3 suffers from the disadvantage that the whole efficient set E(i) is generated at the last step and until this stage we cannot, in general, recognise a single efficient path. This should be contrasted with typical versions of multicriteria simplex methods where we first generate an efficient vertex and then gradually add more points to our tally of efficient vertices until eventually the whole set is built up. This has the advantage that if we fail to complete the whole procedure we still have a useful collection of efficient vertices whereas in our algorithm no useful information is available if we suffer premature termination.

One natural approach to overcoming this disadvantage is to use weighting factors. Thus, we choose a p-vector $\underset{\sim}{w}$ with positive components and solve the scalar network with arc lengths $\underset{\sim}{w}^T \underset{\sim}{c}(i,j)$. It is easy to check that any shortest path from i to t in this network is in E(i). We would then hope that by varying w we could generate all of E(i).

Unfortunately an example in [9] shows that this need not be the case.

A second promising line of approach, suggested in [8], would consider the following scheme. Set

$$\underset{\sim}{v}_o(i) = \infty \qquad \text{for } i \in S \ (i \neq t)$$

$$\underset{\sim}{v}_n(t) = \underset{\sim}{0} \qquad \text{for } n = 0,1,2,\ldots.$$

and for $i \in S$, $(i \neq t)$ and $n = 1,2,\ldots$, let $\underset{\sim}{v}_n(i)$ be any member of

$$\text{eff } \{\underset{\sim}{c}(i,j) + \underset{\sim}{v}_{n-1}(j) \mid j \in \Gamma(i)\}$$

We can readily deduce from the results of the preceding section that all members of $V_N(i)$ and all paths in $F(i)$ can be produced by such a procedure. Unfortunately, as the following example shows, we need not have $\underset{\sim}{v}_N(i) \in V_N(i)$, contrary to assertions by White.[9]

We consider the example in Figure 1. We will take t=7. This gives

$$\underset{\sim}{v}_1(1) = \underset{\sim}{v}_1(2) = \underset{\sim}{v}_1(5) = \infty \qquad \underset{\sim}{v}_1(3) = \underset{\sim}{v}_1(4) = \underset{\sim}{v}_1(6) = \underset{\sim}{v}_1(7) = \begin{pmatrix}0\\0\end{pmatrix}$$

$$\underset{\sim}{v}_2(1) = \infty \quad \underset{\sim}{v}_2(2) = \begin{pmatrix}2\\4\end{pmatrix} \qquad \underset{\sim}{v}_2(3) = \underset{\sim}{v}_2(4) = \ldots. = \underset{\sim}{v}_2(7) = \begin{pmatrix}0\\0\end{pmatrix}$$

$$\underset{\sim}{v}_3(1) = \begin{pmatrix}5\\4\end{pmatrix} \quad \underset{\sim}{v}_3(2) = \begin{pmatrix}2\\4\end{pmatrix} \qquad \underset{\sim}{v}_2(3) = \underset{\sim}{v}_2(4) = \ldots.. = \underset{\sim}{v}_2(7) = \begin{pmatrix}0\\0\end{pmatrix}$$

and $\underset{\sim}{v}_n(i) = \underset{\sim}{v}_3(i)$ for all $i \in S$ and $n = 4,5,\ldots.$

We have exercised choice in obtaining $\underset{\sim}{v}_2(2)$ [from $\begin{pmatrix}2\\4\end{pmatrix}$ and $\begin{pmatrix}3\\3\end{pmatrix}$] and $\underset{\sim}{v}_3(1)$ [from $\begin{pmatrix}3\\5\end{pmatrix}$ and $\begin{pmatrix}5\\4\end{pmatrix}$]. Note that we have also obeyed the additional rules described in [8]. viz. that $\underset{\sim}{v}_n(i) \le \underset{\sim}{v}_{n-1}(i)$ for all $n = 1,\ldots.$ and $i \in S$. Thus $\underset{\sim}{v}_N(1) = \underset{\sim}{v}_3(1)$ but the path $P = 1 \to 2 \to 3 \to 7$ has length $c(P) = \begin{pmatrix}4\\4\end{pmatrix}$ and so $\underset{\sim}{v}_N(1) > c(P)$ which means $\underset{\sim}{v}_N(1) \notin V_N(1)$.

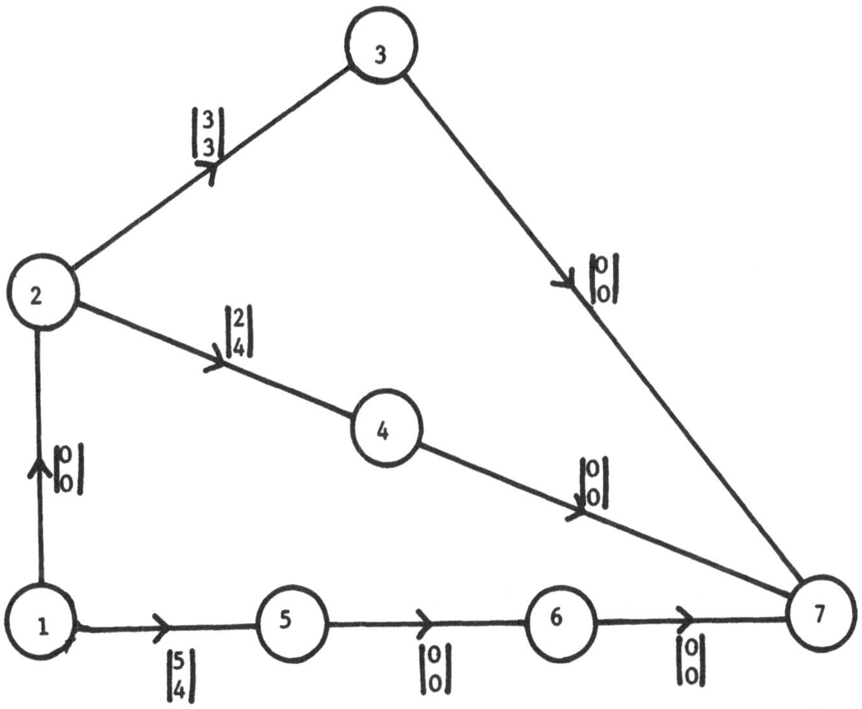

Figure 1

5. References

1. Randolf, P. and Ringeisen, R., Shortest paths through multiple criteria networks, Purdue University preprint, Indiana, 1976.

2. Thuente, D., Two algorithms for shortest paths through multiple criteria networks, Purdue University preprint, Indiana, 1979.

3. Hansen, P., Bicriterion path problems, in Fandel, G. and Gal.T.Eds. *Multiple Criteria Decision Making Theory and Application*, Springer-Verlag, 1980.

4. Climaco, J.C.N., & Martins, E.Q., A bicriterion shortest path algorithm, *E.J.O.R. 11*, 1982, pp. 399-404.

5. Martins, E.Q.V., On a special class of bicriterion path problems, *E.J.O.R. 17*, 1984, pp. 85-94.

6. Martins,E.Q.V., On a multicriteria shortest path problem, *E.J.O.R.16*, 1984, pp. 236-245.

7. White, D.J., The set of efficient solutions for multiple objective shortest path problems, *Computers and O.R. 9*, pp. 101-107.

8. Bellman, R., On a routing problem, *QUART.Appl.Math. 16*, 1958, pp 87-90.

9. White, D.J., *Finite Dynamic Programming*, Wiley, Chichester, England, 1978.

10. Yu, P.L. and Seiford, L., Multistage decision problems with multiple criteria, in Nijkamp, P. and Sprönk, J., Eds., *Multiple Criteria Analysis*, Gower, Aldershot, England, 1981.

11. Hartley, R., Survey of algorithms for vector optimisation problems, in French, S. *et. al., Multi-objective decision making*, Academic Press, New York, 1983.

12. White, D. J. Generalised efficient solutions for sums of sets, *Opus,Res. 28*, 1980, pp. 844-846.

PART 2

MODELLING AND APPLICATIONS

MULTIPLE CRITERIA MATHEMATICAL PROGRAMMING: AN OVERVIEW AND SEVERAL APPROACHES

Stanley Zionts
Management Science and Systems
State University of New York at Buffalo

Introduction

Multiple Criteria Decision Making (MCDM) refers to making decisions in the presence of multiple usually conflicting objectives. Multiple criteria decision problems pervade all that we do and include such public policy tasks as determining a country's policy developing a national energy plan, as well as planning national defense expenditures, in addition to such private enterprise tasks as new product development, pricing decisions, and research project selection. For an individual, the purchase of an automobile or a home exemplifies a multiple criteria problem. Even such routine decisions as the choice of a lunch from a menu, or the assignment of job crews to jobs constitute multiple criteria problems. All have a common thread -- multiple conflicting objectives.

In this study, we discuss some of the important aspects of solving such problems, and present some methods developed for solving multiple criteria mathematical programming problems. We also discuss some applications of the methods.

In multiple criteria decision making there is a decision maker (or makers) who makes the decision, a set of objectives that are to be pursued, and a set of alternatives from which one is to be selected.

GOALS, CRITERIA, OBJECTIVES, ATTRIBUTES, CONSTRAINTS, AND TARGETS: THEIR RELATIONSHIPS

In a decision situation we have goals, criteria, objectives, attributes, constraints, and targets, in addition to decision variables. Although goals, criteria, objectives, and targets have essentially the same dictionary definitions, it is useful to distinguish among them in a decision making context.

Criterion. A criterion is a measure of effectiveness of performance. It is the basis for evaluation. Criteria may be further classified as goals or targets and objectives.

Goals. A goal (synonymous with target) is something that is either achieved or not. For example, increasing sales of a product by at least 10% during one year over the previous is a goal. If a goal cannot be or is unlikely to be achieved, it may be converted to an objective.

Objective. An objective is something to be pursued to its fullest. For example, a business may want to maximize its level of profits or maximize the quality of service provided or minimize customer complaints. An objective generally indicates the direction desired.

Attribute. An attribute is a measure that gives a basis for evaluating whether goals have been met or not given a particular decision. It provides a means for evaluating objectives.

Decision Variable. A decision variable is one of the specific decisions made by a decision maker. For example, the planned production of a given product is a decision variable.

Constraint. A constraint is a limit on attributes and decision variables that may or may not be stated mathematically. For example, that a plant can be operated at most twelve hours per day is a constraint.

Structuring an MCDM Situation

Most problems have, in addition to multiple conflicting objectives a hierarchy of objectives. For example, according to Manheim and Hall[1], the objective for evaluating passenger transportation facilities serving the Northeast Corridor of the U.S. in 1980 was "The Good Life." This superobjective was subdivided into four main objectives:

1. Convenience
2. Safety
3. Aesthetics
4. Economic Considerations

These in turn are divided into subobjectives, and so on forming a hierarchy of objectives.

Some of the objectives, such as economic considerations, have attributes that permit a precise performance measurement. Others, such as aesthetics, are highly subjective. Not wanting to convert the word subjective to a noun, we may, therefore, have a subjective objective. Further, the number of objectives may be large in total.

To adequately represent the objectives, we must choose appropriate attributes. Keeney and Raiffa[2] indicate five characteristics the selected attributes of the objectives should have:

1. Complete: They should cover all aspects of a problem.
2. Operational: They can be meaningfully used in the analysis.
3. Decomposable: They can be broken into parts to simplify the process.
4. Nonredundant: They avoid problems of double counting.
5. Minimal: The number of attributes should be kept small.

I recommend that at most the magic number of about 7 (see Miller[3]) objectives be used. Such a limitation tends to keep a problem within the realm of operationality. What happens if there are more than about 7 objectives? First, use constraints to limit outcomes of objectives about which you are sure or about which you feel comfortable about setting such limits. Since constraints must be satisfied at any price, you should not make constraints "too tight." Further, it is useful to check whether feasible alternatives still exist after adding each constraint or after adding a few constraints. An alternative is to treat some of the objectives as goals or targets. We attempt to satisfy the goals. If we can't, we treat them as objectives. We try to get as close to achieving them as possible. We shall go into the idea of doing this mathematically later. Structuring a problem properly is an art, and there is no prescribed way of setting up objectives, goals and constraints.

A Scenario of Management Decision Making

A scenario of management decision making is generally assumed by most researchers:

1. A decision maker (DM) makes a decision.
2. He chooses from a set of possible decisions.
3. The solution he chooses is optimal.

To criticize the scenario, the decision maker, if an individual (as opposed to a group), seldom makes a decision in a vacuum. He is heavily influenced by others. In some instances groups make decisions. Second, the set of possible decisions is not a given. The set of solutions must be generated. The process of determining the set of alternatives may require considerable effort. Third, what is meant by an optimal solution? Since it is impossible to simultaneously maximize all objectives in determining a solution, a more workable definition is needed. A typical definition of optimality is not particularly workable: <u>An optimal decision is one that maximizes a decision maker's utility (or satisfaction)</u>. In spite of the limitations of the decision scenario, it is widely used; its limitations are hopefully recognized.

Some Mathematical Considerations of Multiple Criteria Decision Making

The general multiple criteria decision making problem may be formulated as follows:

$$\begin{aligned} &\text{"Maximize"} && F(x) \\ &\text{subject to:} && G(x) \leq 0 \end{aligned} \tag{1}$$

where x is the vector of decision variables, and $F(x)$ is the vector of objectives to be "maximized'. In some cases it will be convenient to have an intervening vector y where $F(x) = H(y(x))$. For example, y may be a vector of stochastic objectives which is a function of x . In that case, H would be a vector function of the stochastic objectives.

In some cases F will have some components that are ordinal. Attributes such as quality and convenience of location may only be measurable on an ordinal scale. Further, some objectives may be measured only imperfectly. The word maximize is in quotation marks because maximizing a vector is not a well-defined operation. We shall define it in several ways in what follows.

The constraints $G(x) \leq 0$ are the constraints that define the feasible solution space. They may be stated explicitly and if mathematical be either linear or nonlinear. Alternatively, the alternatives may be stated implicitly by listing them as discrete members of a set. It is frequently convenient to assume that the solutions to the constraints can be used to generate a convex set.

The formulation of the multiple criteria decision making problem (1) is one that I believe includes virtually all of the approaches developed, as well as the various multiple criteria problems. It is clearly too general, because only very specific forms of problem (1) can be solved in practice. A linear version of problem (1) is as follows:

$$\text{"Maximize" } Cx$$
$$\text{subject to: } Ax \leq b$$

$x_j \geq 0$, if needed, may be included in the constraints $Ax \leq b$. This particularization of problem (1) is one on which a substantial amount of study has been made. It is referred to as the <u>mu</u>ltiple <u>o</u>bjective <u>l</u>inear programming problem (MOLP) because it is a linear programming problem with multiple objectives. The following theorem is found in several places in the multiple criteria literature.

Theorem: Maximizing a positive weighted sum of objectives $\lambda'F(=\sum_i \lambda_i F_i)$ over a set of feasible solutions yields a nondominated solution.

The theorem does not say that for every nondominated solution there exists a set of weights for which the nondominated solution maximizes the weighted sum. As we shall see, that need not be the case.

The Objective Functions

Let us now consider the objective functions more carefully. The objective functions may all be assumed to be maximized, without loss of generality, because any objective that is to be minimized can be minimized by maximizing the value of its negative. Accordingly, we shall henceforth refer to objectives to be maximized.

What do we do if we have any goals or targets (as defined earlier)? If they all are simultaneously achievable, we simply add constraints that stipulate the specified value be met and not consider them further. Thus, the achievement of the goals is transformed into an admissible solution satisfying all of the constraints. There is an interesting duality between objectives and constraints, in that the two are closely related.

If the goals are not simultaneously achievable, simply adding constraints as above will lead to no feasible solution to the problem.

What must be done in such a situation is to relax some of the goals, or to change goals to objectives as described earlier: to minimize the difference between the goal and the outcome. The idea is to find a solution that is "close" to the goal.

What do we mean by "maximize?" Unlike unidimensional optimization, we want to simultaneously maximize several objectives. Generally that cannot be done. We may define "maximize" in two ways. From a general perspective, one workable definition of "maximize" is to find all nondominated solutions to a problem.

Definition: Dominance

Solution 1 dominates solution 2 if $F(x_1) \geq F(x_2)$ with strict inequality holding for at least one component of F .

A solution is said to be nondominated if no other solution is at least as good as it in every respect and better than it in at least one respect. The concept seems eminently reasonable. By finding all nondominated solutions, one can presumably reduce the number of alternatives. However, for many problems, the number of nondominated alternatives is still too large to help narrow the choice of alternatives.

There may be some instances where we don't want to eliminate dominated solutions. For example, a dominated solution may be sufficiently close to a nondominated solution that we may decide to make a choice based on some secondary criteria not used in the analysis. We may then very well choose the dominated solution based on the secondary criteria. Alternatively, some of the objectives may not be measurable very precisely. In such a situation we may not want to exclude dominated alternatives from further analysis. As an example of the first type, suppose a prospective automobile purchaser is choosing among cars on the basis of price, economy, sportiness, and comfort. Suppose further that a foreign-made car appears somehow to be the best choice, but that there is a domestically produced automobile that is its equal in all respects except that the price is slightly higher. The decision maker may nonetheless decide to purchase the domestic automobile because of its better availability of spare parts. In the second instance, suppose that the same purchaser is considering two domestically-produced automobiles. We assume as before that the cars are the same for all criteria but one--price. Car A has a lower list price than Car B. However, in the purchase of most automobiles, one can obtain discounts. On haggling with dealers, our purchaser may subsequently find that he can purchase Car B for less than Car A. Hence, if he had excluded Car B because of dominance (on the basis of list price), he would have made a mistake.

The reader may feel that in the first case we should have added spare parts availability to our criteria. Though this could have been done, we may generally use criteria such as this as secondary to resolve close cases. Similarly, it can be argued in the second example that the price variable is transaction price and not list price. Therefore, our selected car is not dominated. Nonetheless, it is difficult to accurately measure transaction price!

At the other end of the spectrum, we define "Maximize" in terms of a utility function: a function of the objectives that gives a scalar measure of performance.

Definition: Utility Function

A utility function is a scalar function $u(F(x))$ such that x_1 is preferred to (is indifferent to) x_2 if and only if $u_1 \geq u_2$. Because of our statement of problem (1), we have at an optimal solution (for any feasible change) $\nabla u \leq 0$ where ∇u is the gradient or the vector of partial derivatives of u with respect to the components of F. What $\nabla u \leq 0$ means is that the utility cannot be increased by moving in any feasible direction. Depending on the method to be considered, we will either estimate a utility function u or approximate it locally. In either case, we will use the function or its approximation to identify a most preferred solution.

A Typology of Multiple Criteria Decision Making Models

Quite naturally, different writers have proposed different decision making typologies. My typology consists of two main dimensions:

1. The nature of outcomes--stochastic versus deterministic.
2. The nature of the alternative generating mechanism--whether the constraints limiting the alternatives are explicit or implicit.

These dimensions are indicated in tabular form in Figure 1. The left-hand column includes the implicit constraint models. When the constraints are nonmathematical (implicit or explicit), the alternatives must be explicit. One of a list of alternatives is then selected. The decision analysis problem is included in the implicit constraint category. When the constraints are mathematical and explicit, then the alternative solutions are implicit and may be infinite in number if the solution space is continuous and consists of more than one solution. Problems in the explicit constraint category are generally regarded as mathematical programming problems involving multiple criteria.

More dimensions may be added to the typology. In addition to implicit constraints versus explicit constraints, and deterministic outcomes versus stochastic outcomes, we can identify other dimensions as well. We may classify the number of decision makers as a dimension: one decision maker versus two or more decision makers. We may classify the number of objectives, the nature of utility functions considered, as well as the number of solutions found (one solution versus all nondominated solutions). I have chosen only two dimensions because they seem to be the most significant factors.

	Implicit Constraints (Explicit Solutions)	Explicit Constraints (Implicit Solutions)
Deterministic Outcomes	Choosing Among Deterministic Discrete Alternatives or Deterministic Decision Analysis	Deterministic Mathematical Programming
Stochastic Outcomes	Stochastic Decision Analysis	Stochastic Mathematical Programming

Figure 1
A Multiple Criteria Decision Method

In our presentation we consider only problems having explicit constraints: mathematical programming problems. We further restrict our consideration to deterministic problems because that is where most of the work has been done. Considerable work has also been done on problems having implicit constraints, both deterministic and stochastic.

In virtually all of the work on multiple criteria decision making, the spirit of the model employed is not necessarily to determine the best decision 1 (though that is desirable!), but to help the decision maker in arriving at his decision. This is what Roy[4] refers to as "decision aid." It is also what Keeney and Raiffa[2] refer to as "getting your head straightened out." Before we consider some of the methods in detail, we present two examples. The first is useful in illustrating some concepts; the second will be used in various forms to illustrate the methods.

Two Examples

Consider the following problem, which we shall refer to as Example 1:

$$\begin{aligned} \text{Maximize} \quad & f_1 = -x_1 + 2x_2 \\ & f_2 = 2x_1 - x_2 \\ \text{subject to:} \quad & x_1 \leq 4 \\ & x_2 \leq 4 \\ & x_1 + x_2 \leq 7 \\ & -x_1 + x_2 \leq 3 \\ & x_1 - x_2 \leq 3 \\ & x_1, x_2 \geq 0 \end{aligned}$$

A plot of the feasible solutions is shown in Figure 2, the maximum solutions indicated (e for f_1 , and b for f_2) for each of the objectives. In that figure, we have also identified all of the feasible extreme point solutions as 0 and a through h . In Figure 3 we have plotted the values of the objective functions for this problem. Each of the feasible solutions in Figure 2 has a corresponding point in Figure 3. For example, solution b represented as $x_1 = 4$ and $x_2 = 1$ has objective function values $f_1 = 12$ and $f_2 = 7$ and is so plotted in Figure 2. The nondominated solutions are shown as the heavy broken line b, c, d, e. An optimal solution presumably will be found along that line, since any point not on that line is either dominated (below and/or to the left) or infeasible (above and/or to the right).

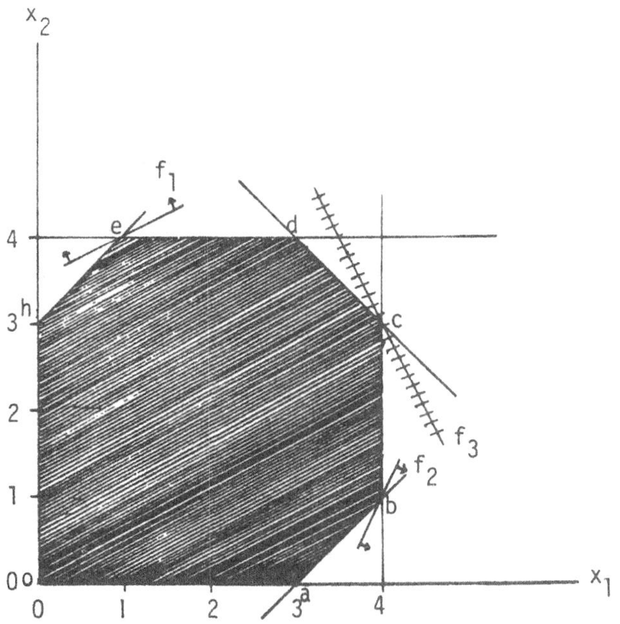

Figure 2
The feasible region of example one and the two objectives f_1 and f_2

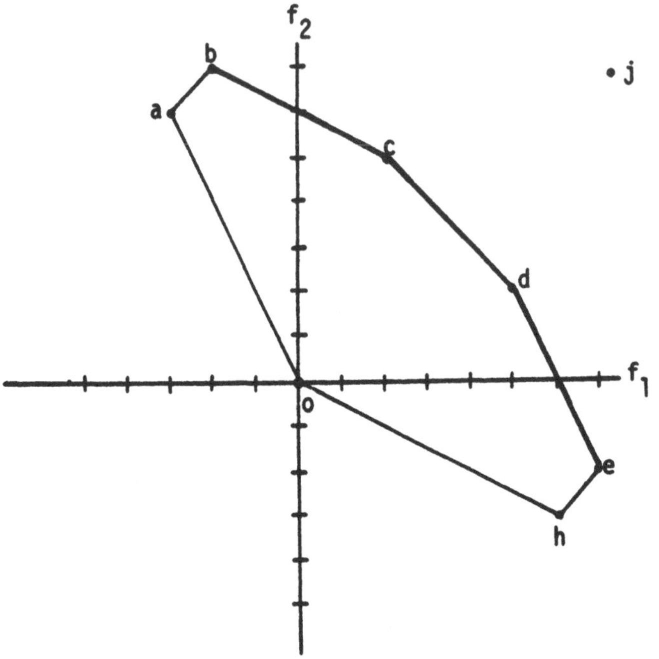

Figure 3
A plot of the solutions of the first example problem in objective function space: in terms of the values of the objective functions.

Since x_1 and x_2 are the decision variables, a graph in terms of x_1 and x_2 (Figure 2) is a graph in decision or activity space. Variables f_1 and f_2 are the objectives; a graph in terms of f_1 and f_2 (Figure 3) is a graph in objective function space. Our example consists of two variables and two objectives. Usually the number of variables is much greater than the number of objectives. We may make the first example more complicated by adding a third objective: $f_3 = 2x_1 + x_2$. See the cross-hatched line in Figure 2. The objective function f_3 is maximized at point c ; the plot of the feasible solutions in decision variable space does not change otherwise. To make a plot in objective function space with three objectives, we would have to add a third dimension to Figure 3. Rather than do that, we first reconsider Figure 3 with two objectives. Denoting as a weighted objective function $\lambda_1 f_1 + \lambda_2 f_2$, we can see that (assuming $\lambda_1 + \lambda_2 = 1$) for $\lambda_1 > 2/3$ solution e is optimal. For $\lambda_1 = 2/3$ both solutions d and e (as well as the solutions on the line between them) are optimal. For $1/2 < \lambda_1 < 2/3$, solution d is optimal. Similarly, for $1/3 < \lambda_1 < 1/2$, solution c is optimal, and for $0 < \lambda_1 < 1/3$, solution b is optimal. Because $\lambda_2 = 1 - \lambda_1$, we could plot the regions along a straight line. Adding a third objective gives us a weighted objective function $\lambda_1 f_1 + \lambda_1 f_2 + \lambda_3 f_3$. Now using the restriction $\lambda_1 + \lambda_2 + \lambda_3 = 1$ or $\lambda_3 = 1 - \lambda_1 - \lambda_2$ we may draw the regions in which each solution is optimal. See Figure 4. The solutions

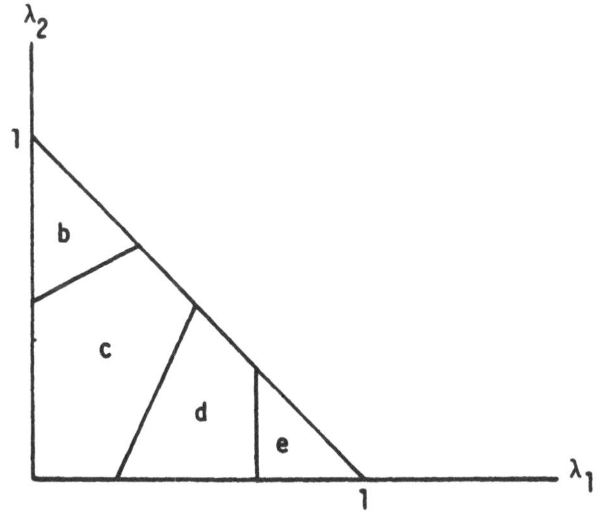

Figure 4
A plot indicating the values of λ_1 and λ_2 $(\lambda_3 = 1 - \lambda_1 - \lambda_2)$

with $\lambda_3 = 0$ (i,e., $\lambda_1 + \lambda_2 = 1$) are still valid; they appear along the line $\lambda_1 + \lambda_2 = 1$. Other solutions are indicated accordingly.

We now consider a more complicated example, Example 2.

$$2x_1 + x_2 + 4x_3 + 3x_4 \leq 60 \quad \text{(slack } x_5)$$

$$3x_1 + 4x_2 + x_3 + 2x_4 \leq 60 \quad \text{(slack } x_6)$$

$$x_1, x_2, x_3, x_4 \geq 0$$

Three objectives are to be maximized:

$$u_1 = 3x_1 + x_2 + 2x_3 + x_4$$

$$u_2 = x_1 - x_2 + 2x_3 + 4x_4$$

$$u_3 = -x_1 + 5x_2 + x_3 + 2x_4$$

The problem has nine basic feasible solutions which are listed below (all omitted variables are zero):

1. $x_1 = 18,\ x_3 = 6,\ u_1 = 66,\ u_2 = 30,\ u_3 = -12$
2. $x_4 = 20,\ x_6 = 20,\ u_1 = 20,\ u_2 = 80,\ u_3 = 40$
3. $x_2 = 15,\ x_5 = 45,\ u_1 = 15,\ u_2 = -15,\ u_3 = 75$
4. $x_2 = 6,\ x_4 = 18,\ u_1 = 24,\ u_2 = 66,\ u_3 = 66$
5. $x_1 = 12,\ x_4 = 12,\ u_1 = 48,\ u_2 = 60,\ u_3 = 12$
6. $x_2 = 12,\ x_3 = 12,\ u_1 = 36,\ u_2 = 12,\ u_3 = 72$
7. $x_3 = 15,\ x_6 = 45,\ u_1 = 30,\ u_2 = 30,\ u_3 = 15$
8. $x_1 = 20,\ x_5 = 20,\ u_1 = 60,\ u_2 = 20,\ u_3 = -20$
9. $x_5 = 60,\ x_6 = 60,\ u_1 = 0,\ u_2 = 0,\ u_3 = 0$

The first six solutions are nondominated, the last three are dominated. Figure 5 indicates which solutions are adjacent extreme point solutions of which other solutions (i.e., they differ by precisely one basic variable).

Solution	is adjacent to Solutions
1	5, 6, 7, 8
2	4, 5, 7, 9
3	4, 6, 8, 9
4	2, 3, 5, 6
5	1, 2, 4, 8
6	1, 3, 4, 7
7	1, 2, 6, 9
8	1, 3, 5, 9
9	2, 3, 7, 8

Figure 5
Adjacency of basic feasible solutions of Example 2.

In order to plot the problem solutions in activity space we need to plot a 4-dimensional graph! More reasonable is plotting the objectives in three dimensions. However, instead we present the plot for the weights λ_1 and λ_2 (and λ_3) as we did for Example 1. See Figure 6.

Any solutions which have a common edge in Figure 6 are adjacent. (See Figure 5.) However, some solutions are adjacent (e.g., 3 and 4), yet do not have a common edge.

Some Naive Methods of Solving Multiple Criteria Mathematical Programming Methods

There are several naive methods for solving multiple criteria mathematical programming problems. They are simple in concept, though generally not very good. Sometimes these naive methods do prove effective, or provide a stepping stone in developing more effective methods. Many of these ideas have led to effective methods. We shall consider only multiple criteria linear programming problems here.

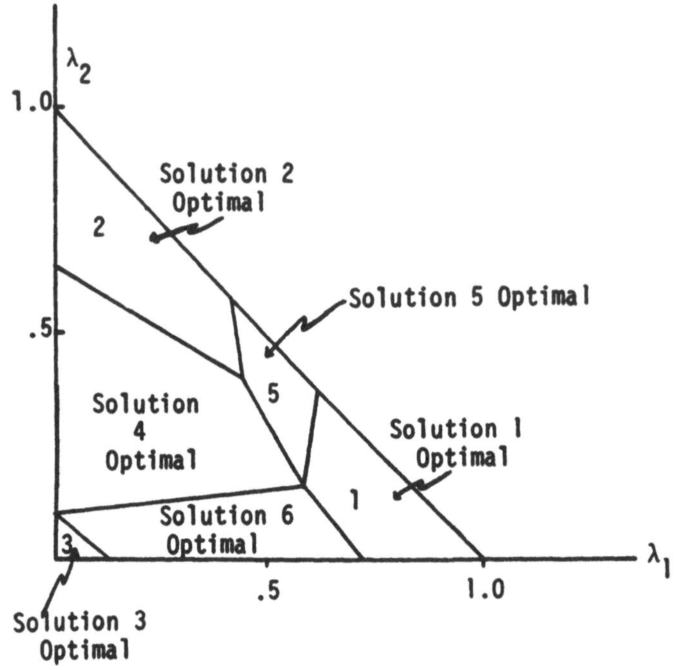

Figure 6
A plot of λ values and the corresponding optimal solutions.

1. Setting Levels of All Objectives

The first of the naive methods to be considered is that of specifying or setting levels of all objectives, and then solving for a feasible solution. The approach is to specify a vector d such that $Cx = d$. The object then is to find a feasible solution to the set of constraints:

$$Cx = d$$

$$Ax \leq b, \; x \geq 0$$

The problem can be solved as a linear programming problem, and there are three possible outcomes as illustrated in Figure 7 for a two-objective problem. The feasible region is indicated. The three possible outcomes are as follows:

a. No feasible solution

b. A dominated solution

c. A nondominated solution.

These are illustrated in Figure 7. If the objectives are set too high, there is no feasible solution (e.g., point a). If the objectives are not set high enough, a feasible solution that is dominated (e.g., solution b) will be found. Almost certainly one of these two outcomes will occur. Only in rare circumstances would simply selecting a vector yield an efficient (or nondominated) solution. Given two points such as

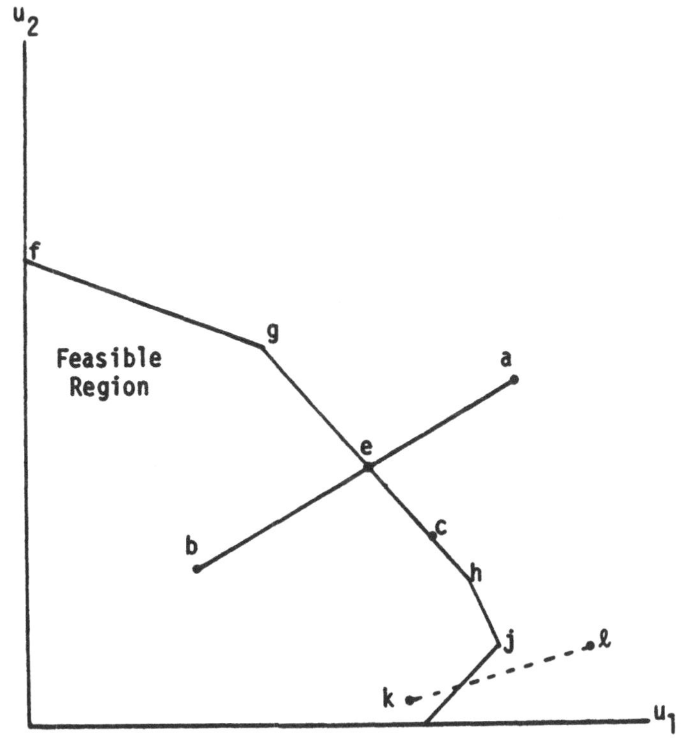

Figure 7
A Graph of a Simple Two Dimensional Example

a and b, we can sometimes use a line search for a nondominated solution on the line segment connecting them (e.g., line segment ab; the nondominated solution would be point e). That this does not necessarily happen is illustrated by feasible point k and infeasible point ℓ; there is no efficient point on the line segment joining them.

Even if we had a method of finding an efficient solution, we would not necessarily know which solution is best. Methods that set levels of all objectives but overcome some of the limitations include goal programming and a method that has been developed by Wierzbicki[5]. These are discussed in later sections. See also the step method (Benayoun *et al*[6].

2. Setting Minimal Levels of All But One Objective

A second naive approach is to set minimum levels for all but one objective and to maximize the remaining objective. Mathematically this amounts to solving a linear programming of the following form:

$$
\begin{aligned}
\text{Maximize} \quad & C_1x \\
\text{subject to} \quad & C_2x \geq d_2 \\
& C_3x \geq d_3 \\
& \vdots \\
& C_px \geq d_p \\
& Ax \leq b, \quad x \geq 0
\end{aligned}
$$

where $d_2,\ldots,d_p$ are the minimum levels of objectives $2,\ldots,p$ and $C_1,C_2,\ldots,C_p$ are the p objective function vectors. We have chosen to maximize the first objective without loss of generality. The result will certainly be a nondominated solution* For our example problem of Figure 2 there are infinitely many solutions along the line segments fg, gh, and hj . Presumably, one (or more) of these solutions is preferred to the others. Which of these solutions is not preferred by the decision maker? That is not clear. A method that employs this approach has been developed by Haimes and Hall[7].

3. Finding All Efficient Extreme Point Solutions

Multiple Objective Linear Programming (MOLP) to find all nondominated or efficient solutions has been widely proposed as another approach. The concept of vector maximum and its early consideration by researchers (see for example, Charnes and Cooper[8]) has been around for a long time. Only in the early 70's was it considered seriously as a computational procedure. Evans and Steuer[9], and Yu and Zeleny[10] generated and solved problems of several sizes to obtain all nondominated extreme point solutions. The results were not good, except for two-objective problems for which parametric programming may be used. Basically, the methods consider the linear programming problem:

*Ralph Steuer has pointed out that solutions to such problems may be in some cases weakly dominated.

$$\begin{aligned} \text{Maximize} \quad & \lambda'Cx \\ \text{subject to:} \quad & Ax = b \\ & x \geq 0 \end{aligned}$$

where the vector of weights $\lambda > 0$. For every nondominated extreme point solution, there exists a convex cone in λ space, that is a cone for which $\lambda'(C_N - C_B{}^{-1}N) \leq 0$, using the usual linear programming notation (C_N and C_B are the nonbasic and basic partitions, respectively, of C, N is the complement of B with respect to A .) The λ-space shown in the various figures is the intersection of all cones with the constraint $\sum \lambda_j = 1$. The methods for finding all nondominated extreme point solutions essentially enumerate the convex cones. The idea was that all efficient solutions could be computed, and the decision maker could choose from them. Since there are in general far too many, the approach is not workable in practice. Steuer's contracting cone method, described in a later section, partially overcomes the problems.

4. Using Weights to Combine Objective Functions

The idea of using weights seems to be an attractive one. It involves averaging or blending the objectives into a composite objective and then maximizing the result. The difficulty is in specifying weights. It is incorrect to say that if the weight for one objective is larger than that of another, that the first objective is more important than the second and <u>vice versa</u>. The weights depend upon the units in which the objectives are measured. For example, equal weights have a rather unequal effect if objectives are to maximize GNP measured in billions of dollars and to maximize the fraction of the population who are above the poverty level as measured by a number between zero and one. The second objective will in that case have virtually no effect. The Zionts-Wallenius method (considered below) extends and uses this approach.

Overcoming the Problems of the Naive Approaches

Several of the naive approaches have appealing characteristics, which no doubt led to their development. To overcome some of the problems with the methods, further development was done on these methods. We now describe the results.

1. Goal Programming

The concept of goal programming, effectively a method for setting all objectives, was introduced by Charnes and Cooper[8], and extended by Ijiri[11] and Lee[12], among others. Goal programming involves the solution of linear programming problems (although other mathematical programming forms such as integer programming have also been formulated in a goal programming context) with several goals or targets. Generally, goal programming assumes a linear constraint set of the (matrix) form $Ax = b, x \geq 0$ where x is the vector of decision variables. Denoting an objective as $c_i x$, there are several possible forms, all of which can be written as $h_i \leq c_i' x \leq u_i$ where h_i is the desired lower bound on objective i, and u_i is the desired upper bound. The bound constraints are not "hard" in that they can be violated. First add variables s_i and t_i and rewrite the bound constraints as

$$c_i'x - x_i \leq u_i , \qquad i = 1,\ldots,p$$

$$c_i'x + t_i \geq h_i , \qquad i = 1,\ldots,p$$

where p is the number of objectives or goals. Now using matrix notation with $c_i = (c_{i1},c_{i2},\ldots,c_{ip})'$, $s = (s_1,s_2,\ldots,s_p)'$, $t = (t_1,t_2,\ldots,t_p)'$, $k = (k_1,k_2,\ldots,k_p)'$, $q = (q_1,q_2,\ldots,g_p)'$, $u = (u_1,u_2,\ldots,u_p)'$ and $h = (h_1,h_2,\ldots,h_p)'$ we wish to

$$\begin{array}{lll} \text{Minimize} & k's + q't & \\ \text{subject to:} & Cx - s & \leq u \\ & Cx \quad + t & \leq h \\ & Ax & = b \\ & x,\ s,\ t & \geq 0 \end{array}$$

where k and q are vectors of weights to measure the violations of the bound constraints. If desired, several different s and t variables may be used for each goal with different values of k and q as well as upper bounds on the s and t variables. The effect of this is to allow for piece-wise linear nonlinear penalties in the failure to achieve goals. As outlined, the relationships yield convex sets. For more information on these nonlinearities as well as nonconvex nonlinearities, see Charnes and Cooper[13].

The bound constraints may be of several different forms. If $u_i = h_i$, the goal is a desired fixed level that is sought. (In that case we need only one goal constraint $c_ix - s_i + t_i = u_i$ $(= h_i)$.) If $u_i > h_i$, the goal is a range. See Figure 8. The penalties may be symmetric or not. If u_i is infinite (or h_i negatively infinite), the corresponding constraint may be omitted. A goal may, therefore, operate as a threshold plus an objective, that is a threshold that is desired, plus an objective that is operable given the threshold is attained. Thus, the formulation possibilities with goal programming are indeed general.

Instead of minimizing a weighted sum of deviations from goals, goal programming may be used to minimize the maximum deviation from a set of goals. This is done by changing the formulation by adding the constraint

$$q_i s_i \leq z$$
$$k_i t_i \leq z$$

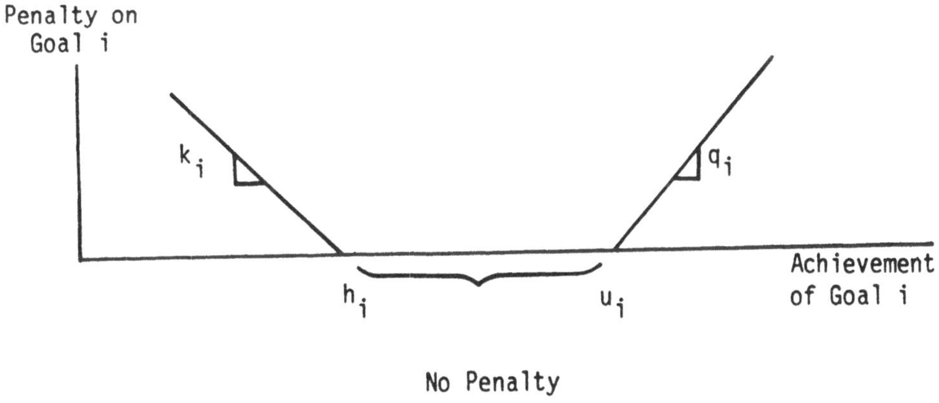

Figure 8
A Goal that is a Range

and changing the objective to minimize z. The effective objective then is to minimize $(\max\{q_i s_i, k_i t_i\})$, the maximum weighted deviation from a goal.

Another variation of goal programming employs preemptive priorities instead of numerical weights. Let some subset of weights have much greater values than another subset of weights so that any finite multiple of the weights of the latter set is always less than any of the weights of the former set. The effect is to first minimize the weighted sums for the highest preemptive priority group. Then constraining that weighted sum to be equal to its minimum value, the next highest preemptive priority group sum is minimized, and so on, for as many preemptive priority groups as there may be.

Where goal programming falls flat is in the selection of the goals as well as the specifications of the weights, that is the vectors k and q. The selection of goals should not be a difficult problem, although it is important for the decision maker to be aware of tradeoffs which face him. The weights must be selected by the user, and goal programming does not have much to say about the choice of weights. About the only device that is offered in terms of weights is preemptive priorities, which we have already considered. Nonetheless, goal programming has been fairly widely used in practice because of the ease of specifying a goal vector, and the ease of understanding what is going on. We consider Example 2 as a goal programming problem, minimizing the absolute sum of (negative) deviations only from the goal (66, 80, 75).

The formulation is as follows:

$$\begin{array}{llr} \text{Minimize} & t_1 + t_2 + t_3 & \\ \text{subject to:} & 3x_1 + x_2 + 2x_3 + x_4 + t_1 & \geq 66 \\ & x_1 - x_2 + 2x_3 + 4x_4 + t_2 & \geq 80 \\ & -x_1 + 5x_2 + x_3 + 2x_4 + t_3 & \geq 75 \\ & 2x_1 + x_2 + 4x_3 + 3x_4 & \leq 60 \\ & 3x_1 + 4x_2 + x_3 + 2x_4 & \leq 60 \\ & x_1, x_2, x_3, x_4, t_1, t_2, t_3 \geq 0 & \end{array}$$

The optimal solution to the above problem is $x_2 = 6$, $x_4 = 18$, $t_1 = 42$, $t_2 = 14$, $t_3 = 9$ (all other variables are zero), or $u_1 = 24$, $u_2 = 66$, $u_3 = 66$. Changing the objective function to minimize $3t_1 + t_2 + t_3$ changes the solution to $u_1 = 66$, $u_2 = 30$, $u_3 = -12$, to illustrate another set of weights.

If we now add to the formulation $t_1 \leq z$, $t_2 \leq z$, and $t_3 \leq z$ and change the objective to minimize z we have an example of minimizing the maximum deviation from each goal. We obtain the solution $x_1 = 4.86$, $x_2 = 5.014$, $x_3 = 2.88$, $x_4 = 11.24$, $t_1 = t_2 = t_3 = t_4 = z = 29.40$, or $u_1 = 36.6$, $u_2 = 50.6$, and $u_3 = 45.6$.

We now illustrate the use of preemptive priorities. Let us assume that our first priority is to get u_1 to 50, our second priority is to get u_2 to 50, and our third priority is to get u_3 to 50.

We formulate the problem as follows:

$$\begin{array}{llll}
\text{Minimize} & P_1t_1 + P_2t_2 + P_3t_3 & & (P_1 \gg P_2 \gg P_3) \\
& 3x_1 + x_2 + 2x_3 + x_4 + t_1 & \geq 50 & (u_1 \geq 50) \\
& x_1 - x_2 + 2x_3 + 4x_4 + t_2 & \geq 50 & (u_1 \geq 50) \\
& -x_1 + 5x_2 + x_3 + 2x_4 + t_3 & \geq 50 & (u_2 \geq 50) \\
& 2x_1 + x_2 + 4x_3 + 3x_4 & \leq 60 & \\
& 3x_1 + 4x_2 + x_3 + 2x_4 & \leq 60 & \\
& x_1, \ldots, x_4, t_1, t_2, t_3 & \geq 0 &
\end{array}$$

The optimal solution to the problem is $u_1 = 50$, $u_2 = 50$, u_3 13.65 ($x_1 = 11.88$, $x_2 = 1.18$, $x_3 = 2.24$, and $x_4 = 8.71$) .

2. Scalarizing Functions and the Method of Wierzbicki

Wierzbicki[5] has developed a method which may be thought of as a method for setting levels of all objectives. It assumes that all objectives are to be maximized, and employs a scalarizing function to find an efficient solution. Referring to our naive version, the chosen levels of objectives are almost certainly infeasible or dominated. The scalarizing method or reference point approach, as it also is called (see Kallio, Lewandowski, and Orchard-Hays[14]), find the closest efficient solution to the chosen point. It is intended to be used in a simulation-type mode by the decision maker.

Although there are a wide variety of scalarization functions that could be used, one that seems quite effective is one which can be represented in a linear programming context. Let u_i be the target level for objective i . The objective is to maximize

$$\left\{ \min \left\{ \rho \min_i \{C_i x - u_i\} \ , \sum_i (C_i x - u_i) \right\} \ + \varepsilon(C_i x - u_i) \right\}$$

where the parameter $\rho \geq p$ the number of objectives and ε is a nonnegative vector. This objective function is achieved by a similar representation to that in goal programming for the minimization of the maximum deviation from a set of goals. Here, however, we maximize the minimum of (1) a constant times the minimum overachievement of a goal, and (2) the sum of overachievements of goals, averaged together with a weighted overachievement of goals. As in the case of goal programming, the function and parameters are somewhat arbitrary. However, the purpose of this method is to be used as an efficient solution generator, one that can be used to generate a sequence of efficient solution points. It is rather similar to goal programming and has been programmed to solve problems having as many as 99 objectives with as many as 1000 constraints.

3. Steuer's Contracting Cone Method

Steuer's Contracting Cone Method (Steuer and Schuler[15] and Steuer and Wallace[16]. See also Steuer[17].) is a refinement to the generation of all nondominated solutions that generates only a relatively small number of nondominated extreme point solutions. It does this by selecting a convex cone in λ space that is large initially and includes sets of weights corresponding to many nondominated extreme point solutions. Rather than generating all of them, however, he generates only a very small number of extreme point solutions, and questions the decision maker regarding their relative attractiveness. He then uses the responses to contract the cone. When the cone becomes sufficiently small, the method generates all of the nondominated extreme point solutions in the cone for final consideration by the decision maker.

Assuming that there are p objectives, Steuer's method generates 2p + 1 trial solutions each time. The vectors generated* are:

	Values in General	Initial Values
λ_1	= the first extreme vector	$(1,0,0,\dots,0)$
λ_2	= the second extreme vector	$(0,1,0,\dots,0)$
$\vdots$	$\vdots \quad \vdots$	
λ_p	= the p the extreme vector	$= (0,0,0,\dots,1)$
λ_{p+1}	$= 1/p\ (\lambda_1 + \lambda_2 + \dots + \lambda_p)$	$= (1/p,1/p,1/p,\dots,1/p)$
λ_{p+2}	$= (\lambda_2 + \lambda_3 + \dots + \lambda_{p+1})/p$	$= (1/p^2,r,r,r,\dots,r)$
λ_{p+3}	$= (\lambda_1 + \lambda_3 + \lambda_4 + \dots + \lambda_p + \lambda_{p+1})/p$	$= (r,1/p^2,r,r,\dots,r)$
λ_{p+4}	$= (\lambda_1 + \lambda_2 + \lambda_4 + \dots + \lambda_p + \lambda_{p+1})/p$	$= (r,r,1/p^2,r,\dots,r)$
$\vdots$		
λ_{2p+1}	$= (\lambda_1 + \lambda_2 + \dots + \lambda_{p-1} + \lambda_{p+1})/p$	$= (r,r,r,\dots,r,1/p^2)$
	where	$r = (p + 1)/p^2$

The first p vectors are the extreme vectors of the cone, the p + 1st is the mean or center of gravity of the first p vectors, and each of the others is the mean or center of gravity of p - 1 extreme vectors and the p + 1st vector. For each of the weight vectors λ, a linear

*Instead of using zeros in the vector, we use some sufficiently small positive numbers. However, for simplicity of presentation we use zeros here.

programming problem is solved maximizing $\lambda'Cx$, and the decision maker is presented with the $2p + 1$ solutions. He is asked to choose which of the solutions he likes most, or if he is ready to look at all of the extreme point solutions in the cone. In the latter case, all extreme point solutions in the cone are found and presented to the decision maker for a final choice. Otherwise, the cone is contracted about the selected extreme point solution. The first p vectors (the extreme vectors) for the next iteration are the vectors corresponding to the chosen solution, say λ_q , and the average of that vector with each of the (first) p extreme vectors from the previous iteration.

$$\lambda'_i(i=1,\dots,p) = .5\lambda_i + .5\lambda_q$$

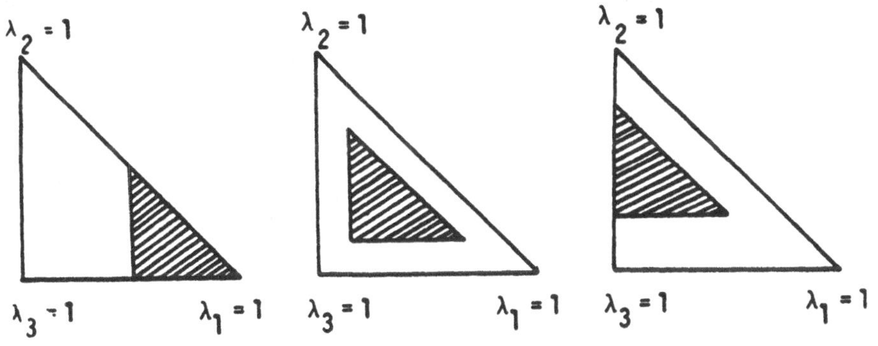

Figure 9
An illustration of the three cases of contracting cones.

Case a The solution corresponding to $\lambda_1 = 1$, $\lambda_j = 0\ (j \neq 1)$, is preferred.

Case b The solution corresponding to $\lambda_j = 1/p$, $j = 1,\dots,p$, is preferred.

Case c The solution corresponding to $\lambda_1 = 1/p^2$, $\lambda_j = (p + 1)p^2$, $j \neq 1$ is preferred.

The prime indicates the new trial weights. The remaining $p + 1$ vectors are found from $\lambda'_1, \ldots, \lambda'_p$ as λ_{p+i}, $i = 1, \ldots, p + 1$ are found from $\lambda_1, \ldots, \lambda_p$.

The process is repeated until the decision maker asks for all efficient solutions defined in a cone to make a final decision. The effect of contracting the cone is to reduce the volume of the cone to $(1/2)^p$ of what it was prior to the contraction. This fraction could be adjusted to a larger or smaller fraction, if desired. To illustrate how the cone contracts as we've described, consider a three objective problem. Figure 9 illustrates such a cone section with $\lambda_1 + \lambda_2 + \lambda_3$. If we contract the cone about one of the original p extreme vectors (λ_1) , we have the diagram shown in Figure 9a. If we contract the cone about the center of gravity (the mean -- λ_4), we have the diagram shown in Figure 9b. Finally, if we contract the cone about one of the off-center solutions (λ_5) , we have the diagram shown in Figure 9c.

The procedure is appealing, but heuristic in nature. It does not always find the optimal solution. Consider the following problem, contrived from Example 2.

$$\begin{aligned} \text{Maximize} \quad u_1 &= 66x_1 + 20x_2 + 15x_3 + 24x_4 + 29.6x_5 + 36x_6 \\ u_2 &= 30x_1 + 80x_2 - 15x_3 + 66x_4 + 69.6x_5 + 12x_6 \\ u_3 &= -12x_1 + 40x_2 + 75x_3 + 66x_4 + 29.6x_5 + 72x_6 \end{aligned}$$

$$\begin{aligned} \text{Subject to} \quad & x_1 + x_2 + x_3 + x_4 + x_5 + x_6 \le 1 \\ & x_1, \ldots, x_6 \ge 0 \end{aligned}$$

Suppose that the true set of weights is $\lambda_1 = .5205$, $\lambda_2 = .4$, $\lambda_3 = .0795$. The optimal solution is then $x_5 = 1$, all other x's zero. Though we will not show it, Steuer's method will almost certainly identify solution $x_2 = 1$, all other x's zero as optimal.

The method has the capability of generating all of the efficient solutions in a convex cone of weights λ. It seems attractive because the cone can be contracted until it is as small as the user wishes. The disadvantages are that the decision maker must choose among as many as $2p + 1$ alternatives each iteration, as opposed to a smaller number. As shown in he contrived example, the optimal solution is not necessarily obtained.

Using example 2, (see Figure 6) we assume a "true" set of weights of $\lambda_1 = .58$, $\lambda_2 = .21$, and $\lambda_3 = .21$; solution 5 is optimal. Steuer's first set of weights and their solutions are as follows:

λ_1	λ_2	λ_3	Solution
1	0	0	1
0	1	0	2
0	0	1	3
1/3	2/3	1/3	4
1/9	4/9	4/9	4
4/9	1/9	4/9	6
4/9	4/9	1/9	5

The cone will be contracted in the region of solution 5, and will certainly find the optimal solution, particularly since all feasible optimal extreme point solutions have been already listed above. Regarding performance, Steuer and Schuler (1976) report favorable experience in applications to a forestry management problem.

4. The Zionts-Wallenius Method

A method for multiple objective linear programming which uses weights is one developed by Zionts and Wallenius[18,19]. In that framework a numerical weight (arbitrary initially) is chosen for each objective. Then each objective is multiplied by its weight, and all of the weighted objectives are then summed. The resulting composite objective is a proxy for a utility function. (The manager need not be aware of the combination process.) Using the composite objective, solve the corresponding linear programming problem. The solution to that problem, an efficient solution, is presented to the decision maker in terms of the levels of each objective achieved. Then the decision maker is offered some trades from that solution, again only in terms of the marginal changes to the objectives. The trades take the form, "Are you willing to reduce objective 1 by so much in return for an increase in objective 2 by a certain amount, an increase in objective 3 by a certain amount, and so on?" The decision maker is asked to respond either yes, no, or I don't know to the proposed trade. The method then develops a new set of weights consistent with the responses obtained, and a corresponding new solution. The process is then repeated, until a best solution is found.

The above version of the method is valid for linear utility functions. However, the method is extended to allow for the maximization of a general but unspecified concave function of objectives. The changes to the method from that described above are modest. First, where possible the trades are presented in terms of scenarios, e.g., "Which do you prefer, alternative A or alternative B?" Second, each new nondominated extreme point solution to the problem is compared with the old, and either the new solution, or one preferred to the old one is used for the next iteration. Finally, the procedure terminates with a neighborhood that contains the optimal solution. Experience with the method has been good. With as many as seven objectives on moderate linear programming problems (about 300 constraints) the maximum number of solutions is about ten, and the maximum number of questions is under 100.

We describe the general concave (GC) version in some detail. The linear problem form may, of course, be solved as a special case, though the GC method does not reduce to the linear method in that case.

We repeat the formulation of the problem for convenience.

$$\text{Maximize} \quad g(Cx)$$
$$\text{subject to:} \quad Ax \leq b, \quad x \geq 0$$

The underlying concave utility function g is assumed to have continuous first derivatives. We present the algorithm as a sequence of steps.

1. Choose an arbitrary vector of weights, $\lambda > 0$.

2. Solve the linear programming problem

$$\text{Maximize} \quad \lambda' Cx$$
$$\text{subject to:} \quad Ax \leq b, \quad x \geq 0$$

   The result is a nondominated extreme point solution x^* . If this is the first time through this step, go to step 3. Otherwise, ask whether solution x^* is preferred to the old

x^* solution. If yes, discard the old solution and go to step 3. If no, replace x^* by x^0 and go to step 3.

3. Find all adjacent efficient extreme point solutions to x^* consistent with prior responses. If there are none, drop the oldest set of responses and repeat step 3. Otherwise go to step 4.
4. (This step is simplified over what is used. See Zionts and Wallenius[19].) Ask the decision maker (DM) to choose between x^* and an adjacent efficient extreme point solution. Do not repeat any questions previously asked. If the objective function values of the solutions are too close, or if x^* was preferred to an adjacent solution, ask the decision maker about the tradeoffs between the two solutions. The DM may indicate which solution he prefers, or indicate that he cannot choose between the two. If he prefers no alternatives or tradeoff go to step 5. Otherwise mark a solution preferred to x^* as x^0 and go to step 6.
5. If all previous responses have been deleted, stop; if the decision maker does not like any tradeoffs from x^*, the optimal solution is x^*. Otherwise, to find the optimal solution, the method terminates and a search method (not part of this method) must be used to search the facets. If the procedure does not stop in which case previous responses have not been deleted, delete the oldest set of responses and go to step 3.

6. Find a set of weights $\lambda > 0$ consistent with all previous responses. If there is no feasible set, delete the oldest response and repeat step 6. When a feasible set of weights is found, go to step 2.

To find the adjacent efficient extreme points in step 3, consider the tradeoffs offered $(w_{1j},\ldots,w_{pj})$ by moving to the adjacent extreme point solution j . Then, consider the following linear programming problem:

$$\text{Maximize} \quad \sum_{i=1}^{p} w_{ik}\lambda_i$$

$$\text{subject to} \quad \sum_{i=1}^{p} w_{ij}\lambda_i \leq 0 \qquad j \in N\,, \quad j \neq k \qquad \text{(A)}$$

$$\qquad\qquad\qquad {}_i \geq 0 \qquad\qquad i = 1,\ldots,p$$

where N is the set of nonbasic variables corresponding to solution on x_* . No convex combination of tradeoffs dominates the null vector, for otherwise solutions x^* would not be efficient.

<u>Definition</u>: Given two efficient extreme point solutions x_a and x^*, solution x_a is an adjacent efficient extreme point solution of x^* if and only if all convex combinations of x^* and x_a are efficient solutions.

<u>Theorem</u>: The optimal solution to problem (A) is zero if and only if solution k offering the tradeoff vector $w_{1k},\ldots,w_{pk}$ is not an efficient vector of the set of vectors w_j, $e \in N$. (For a proof see Zionts and Wallenius[18].

Corollary: If problem (A) has a positive infinite solution, then solution k offering the tradeoff vector $w_{1k},\ldots,w_{pk}$ is an efficient vector of the set of vectors w_j, $j \in N$.

The method does not explicitly solve problem (A) for every value of k . What it does is to choose one value of k at a time and to solve (A) for that value of k . At each iteration, a sequence of tests for other values of k are made which in general eliminate solving problems for subsequent values of k .

As an example of the method, we consider Example 2. As a "true" set of weights we use $\lambda_1 = .58$, $\lambda_2 = .21$, and $\lambda_3 = .21$. Our solution procedure begins with $\lambda_1 = \lambda_2 = \lambda_3 = 1/3$. Refer to Figure 6 for further insight. The initial solution is solution 4. First the decision maker is asked to compare solutions 4 and 2, he should prefer 4. Considering 4 versus 5, he should prefer 5. Considering 4 versus 6, he should prefer 4. A consistent set of weights is $\lambda_1 = .818$, $\lambda_2 = .182$, $\lambda_3 = 0$, and the new solution* is solution 1. The decision maker is asked to choose between 1 and 5; he should prefer 5. A set of consistent weights is .594, .160, .246 . They yield solution 5. A final question is asked: between 5 and 2. Since he should prefer 5, there are no further questions to ask; solution 5 is optimal.

The Zionts-Wallenius method is extended to integer programming in Zionts[20], which is implemented and tested in Villareal[21]. See also

*We don't use zero weights: λ_3 would be equal to some arbitrary small positive number.

Villareal, Karwan, and Zionts[22,23] and Karwan, Zionts, and Villareal[24]. Other methods of multiple criteria integer programming are summarized in Zionts[25].

The Zionts-Wallenius method has been used by several organizations and has met with success. For example, Wallenius, Wallenius, and Vartia[26] describe an application to macroeconomic planning for the Government of Finland. They used an input-output model of the Finnish economy with four objectives chosen by the Finnish Economic Council chaired by the Prime Minister. The objectives were:

1. the percentage change in gross domestic product
2. unemployment
3. the rate of inflation as measured by consumer prices
4. the balance of trade.

They first tried using the Geoffrion, Dyer, and Feinberg[27] approach using an improvement prescribed by Dyer[28]. Although the method worked, the users found the estimation of the marginal rates of substitution difficult. Then the Zionts-Wallenius method was used. Results were obtained that were quite satisfactory.

One criticism of the Zionts-Wallenius approach is that at termination we may not always have an optimal solution. However, the termination of the procedure indicates when this does occur. In such instances, we will have an extreme point solution that is preferred to all adjacent efficient extreme point solutions. A search procedure will then have to be used to find the optimal. See, for example, Deshpande[29].

5. The Geoffrion, Dyer, and Feinberg Method

The next mathematical programming method to be discussed, that of Geoffrion, Dyer, and Feinberg[27], is in the spirit of a weighting method. However, it is a gradient type method which allows for a nonlinear problem. The method begins with a decision that satisfies all of the constraints. Then information is elicited from the decision maker indicating how he would like to alter the initial levels of the various objectives. More specifically, he is asked to indicate how much of a reference criterion he is willing to give up in order to gain a fixed amount on one of the other criteria. The responses are elicited for every criterion except the reference criterion.

To illustrate, suppose that one has three objectives:

1. to maximize return on investment;
2. to maximize growth in sales;
3. to minimize borrowing.

Given a starting feasible solution and taking return on investment as our reference criterion, the decision maker would be asked two questions to consider from that solution:

1. What percentage growth in sales must you gain in order to give up a 1% return on investment?
2. What decrease in borrowing must you achieve in order to give up a 1% return on investment?

His responses can be used to determine the direction of change in objectives most desired.

That direction is then used as an objective function to be maximized, and the solution (the new solution) maximizing the objective is found. Then a one-dimensional search is conducted with the decision maker from the previous solution to the new solution. The decision maker is asked in a systematic manner to choose the best decision along that direction. Using the best decision as a new starting point, a new direction is elicited from the decision maker as above and the process is repeated until the decision maker is satisfied with the solution.

We now give an example of the Geoffrion, Dyer, Feinberg method. If we were to assume a linear utility function, and provide correct tradeoff information, the method requires only one iteration. If we assume a nonlinear utility function or consider a linear utility function and do not provide correct tradeoff information, more iterations are required. We choose to assume a nonlinear utility function. We use example two; our utility function is

$$\text{Maximize} \quad U = -(u_1 - 66)^2 = (u_2 - 80)^2 - (u_3 - 75)^2$$

The constraints are as before.

We start with the solution $u_1 = u_2 = u_3 = x_1 = x_2 = x_3 = x_4 = 0$ ("true" objective function value -16,381). The partial derivatives are $\frac{\partial U}{\partial u_1} = -2(u_1 - 66)$, $\frac{\partial U}{\partial u_2} = -2(u_2 - 80)$, and $\frac{\partial U}{\partial u_3} = -2(u_3 - 75)$. For the initial solution the vector of partial derivatives is 132, 160, 15, normalized as .299, .362, .339 . We solve the linear programming problem using this set of weights to combine objectives. From Figure 6 (p. 244) we see the solution with that set of weights is solution 4 (24 66 66) . We then choose the best solution on the line segment between (0 0 0) and (24 66 66) which is (24 66 66) (with true objective function value -2041). We find this by searching along the line segment between the two solutions. The new normalized objective function vector at (24 66 66) is .646 .215 .138, and the solution for that set of weights is solution 5 (48 60 12) . The maximum solution on the line segment between the two solutions. The new normalized objective function vector at (24 66 66) is .646 .215 .138, and the solution for that set of weights is solution 5 (48 60 12). The maximum solution on the line segment between (26.6 63.8 61.2) and (48 60 12) is (27.4 63.7 59.5) (with true objective function value -1999). At this point we begin to alternate between maximizing solutions four and five until the solution converges. The first few solutions and the optimum are summarized in Table 1. The optimal solution is approximately $x_1 = 1.5$, $x_2 = 5.25$, $x_4 = 17.25$ with objective function values (27.0 65.25 59.25) .

An application of the method to the operation of an academic department on a university campus is described based on data from the 1970-1971 operations of the Graduate School of Management, University of

TABLE 1

	Solution			Objective Function Value	Maximizing Solution		
1	0	0	0	-16,381	24	66	66
2	24	66	66	- 2,041	66	30	-12
3	26.6	63.8	61.2	- 2,005	48	60	66
4	27.4	63.7	59.5	- 1,999	24	66	66
5	27.0	64.0	60.3	- 1,993	48	60	12
6	27.34	63.94	59.52	- 2,992.15	24	66	66
7	27.092	64.093	60.002	- 1,991.8	48	60	12
⋮							⋮
	27.0	65.25	59.25	- 1,986.6	24	66	66
					48	60	12

California, Los Angeles. A linear programming model of the problem was developed and used to formulate annual departmental operating plans. Six criteria for evaluation were stated, including number of course sections offered at various levels, the level of teaching assistance used, and faculty involvement in various nonteaching activities. The decision variables under the control of the department were the number of course sections offered at different levels, the number of regular and temporary faculty hired, and the number of faculty released from teaching. The starting point for the analysis was the previous year's operating position, and the resulting solution suggested an important reallocation of faculty effort from teaching to other activities. The method was used without significant difficulty, and the results were adopted by the department.

The problems in the method are the evaluation of the gradient and the choice of a solution along a line segment in the search procedure.

CONCLUSION

This presentation was designed as an introduction to the multiple criteria decision problem, with a representation to some of the methods that have been developed to solve the multiple objective mathematical programming problem. Our treatment was, of necessity, brief. For further information on the methods presented refer to the references. For additional information on these methods and new developments in the field, refer to the various journals in management science. We have mentioned some applications of the methods. The methods are proving useful in helping people to solve multiple criteria decision problems.

REFERENCES

1. Manheim, M. L. and Hall, F., Abstract representation of goals: a method for making decisions in complex problems, in *Transportation: A Service*, Proceedings of the Sesquicentennial Forum, New York Academy of Sciences American Society of Mechanical Engineers, New York, 1967.

2. Keeney, R. L. and Raiffa, H., *Decisions with Multiple Objectives Preferences and Value Tradeoffs*, John Wiley and Sons, New York, 1976.

3. Miller, G., The magical number seven plus or minus two: some limits on our capacity for processing information, *Psychological Review*, 63, 1956, 81.

4. Roy, B., Partial preference analysis and decision aid: the fuzzy criterion concept, in Bell, D. E., Keeney, R. L. and Raiffa, H. eds., *Conflicting Objectives in Decisions*, International Series on Applied Systems Analysis, John Wiley and Sons, 1977, 442 pp.

5. Wierzbicki, A. P., The use of reference objectives in multiobjective optimization, Working Paper 79-66, International Institute for Applied Systems Analysis, Laxemburg, Austria, 1979.

6. Benayoun, R., de Montgolfier, J., Tergny, J., and Larichev, O., Linear programming with multiple objective functions: step method (STEM), *Mathematical Programming*, 1, 1971, 366.

7. Haimes, Y. Y., and Hall, W. A., Multiobjectives in water resources systems analysis: the surrogate worth trade off method, *Water Resources Research*, 10, 1974, 615.

8. Charnes, A. and Cooper, W. W., *Management Models and Industrial Applications of Linear Programming*, John Wiley and Sons, New York, 1961.

9. Evans, J. P. and Steuer, R. E., Generating efficient extreme points in linear multiple objective programming: two algorithms and computing experience, in Cochrane and Zeleny, *Multiple Criteria Decision Making*, University of South Carolina Press, 1973.

10. Yu, P. L. and Zeleny, M., The set of all nondominated solutions in the linear cases and a multicriteria simplex method, *Journal of Mathematical Analysis and Applications*, 49, 1975, 430.

11. Ijiri, Y., *Management Goals and Accounting for Control*, North-Holland Publishing Co., Amsterdam, and Rand McNally, Chicago, 1965.

12. Lee, S. M., *Goal Programming for Decision Analysis*, Auerbach, Philadelphia, 1972.

13. Charnes, A. and Cooper W. W., Goal programming and multiple objective optimization - Part 1, *European Journal of Operations Research*, 1, 1977, 39.

14. Kallio, M., Lewandowski, A., and Orchard-Hays, W., An implementation of the reference point approach for multiobjective optimization, Working Paper No. 80-35, International Institute for Applied Systems Analysis, Laxemburg, Austria, 1980.

15. Steuer, R. E., and Schuler, A. T., An interactive multiple objective linear programming approach to a problem in forest management, Working Paper No. BA2, College of Business and Economics, University of Kentucky, 1976.

16. Steuer, R. E., and Wallace, M. J., Jr., An interactive multiple objective wage and salary administration procedure, in Lee, S. M., and Thorp, C. D., Jr., eds., *Personnel Management: A Computer-Based System*, Petrocelli, New York, 1978, 159.

17. Steuer, R. E., Multiple objective linear programming with interval criterion weights, *Management Science*, 23, 1977, 305.

18. Zionts, S. and Wallenius, J., An interactive programming method for solving the multiple criteria problem, *Management Science*, 22, 1976, 652.

19. Zionts, S. and Wallenius, J., An interactive multiple objective linear programming method for a class of underlying nonlinear utility functions, *Management Science*, 29, 1983, 519.

20. Zionts, S., Integer linear programming with multiple objectives, *Annals of Discrete Mathematics*, 1, 1977, 551.

21. Villareal, B., *Multicriteria Integer Linear Programming*, Doctoral Dissertation, Department of Industrial Engineering, State University of New York at Buffalo, 1979.

22. Villareal, B., Karwan, M. H., and Zionts, S., An interactive branch and bound procedure for multicriterion integer linear programming, in Fandel, G. and Gal, T., eds., (1980), *Multiple Criteria Decision Making – Theory and Application* Proceedings, Springer-Verlag, Berlin, 1979.

23. Villareal, B., Karwan, M. H., and Zionts, S., A branch and bound approach to interactive multicriteria integer linear programming, paper presented at Joint National Meeting TIMS/ORSA, Washington, D.C., 1980.

24. Karwan, M. H., Zionts, S., and Villareal, B., An improved interactive multicriteria integer programming algorithm, Working Paper No. 530, School of Management, State University of New York at Buffalo, 1983.

25. Zionts, S., A survey of multiple criteria integer programming methods, *The Annals of Discrete Mathematics*, 5, 1979, 389. Reprinted in Hammer, P. L., Johnson, E. L. and Korte, B. H., eds., *Discrete Optimization II*, North-Holland, Amsterdam, 1979.

26. Wallenius, H., Wallenius, J., and Vartia, P., An approach to solving multiple criteria macroeconomic policy problems and an application, *Management Science*, 24, 1978, 1021.

27. Geoffrion, A. M., Dyer, J. S. and Feinberg, A., An interactive approach for multicriterion optimization with an application to the operation of an academic department, *Management Science*, 19, 1972, 357.

28. Dyer, J., A time-sharing computer program for the solution of the multiple criteria problem, *Management Science*, 19, 1973, 349.

29. Deshpande, D., Investigations in multiple objective linear programming-theory and an application, Unpublished Doctoral Dissertation, School of Management, State University of New York at Buffalo, 1980.

AIDS FOR DECISION MAKING WITH CONFLICTING OBJECTIVES

Elemer E. Rosinger
Department of Mathematics
University of Pretoria

ABSTRACT

A succession of *three* man-machine interactive decision aids is presented with increasing *user-friendliness* what the communication between the decision maker and the respective programs is concerned. This increase in user-friendliness is not brought about on the account of other useful features of the decision aids, such as ease in finding the optimal solutions, fidelity in modelling the conflict between various objectives, confidence in the solutions obtained and efficience of the decision maker's participation. Indeed, all the three decision aids continue to exhibit the later four features more or less to the same extent.

0. INTRODUCTION

One of the essential features of decision making processes in management, economics, etc. is that the feasible decisions have to be judged simultaneously from the point of view of *several* different and usually *conflicting* objectives. Within the sequel, we shall address the case where *one single* decision maker (DM), for instance, a top executive, is confronted with such a situation of conflicting objectives.

Historically two well known methods for handling such situations exist. The first one, called the *method of priorities*, consists of listing the various objectives according to a certain order of importance or priority and then judging each feasible decision starting with the objective on the top of the priority list. The feasible decisions which pass this test are then judged according to the second objective on the priority list and so on. A more recent method, which became widely used more or less simultaneiously with the emergence of economics as an independent discipline, is the *method of weighted objectives* which among others, assumes cardinal objectives, i.e., objectives which can be measured according to some units and their measures can be expressed by numbers. The disadvantages of these two traditional methods of dealing with conflicting objectives have been repeatedly experienced in the context of various applications and an early attempt to go beyond these methods was suggested by Pareto who introduced the *solution concept* of *nondominated decisions* which, with the help of a suitable additional analysis, such as for instance a trade-off between various marginal utilities, was supposed to lead to an optimal decision.

In order to have a clearer insight into the difficulties and the corresponding problems which arise in the context of decision processes with conflicting objectives, we shall analyze a model of such decision processes which is sufficiently general so as to include large classes of relevant particular cases. In fact, the only *limitation* of the model is that it assumes *cardinal objectives*. Therefore, suppose given a set X of feasible decisions and $m \geq 2$ cardinal objectives

$$f_1, \ldots, f_m : X \to R^1 \tag{1}$$

which have to be maximized simultaneously. As is known, the basic difficulty which arises in the case of $m \geq 2$ objectives is that if we try to maximize each objective f_j independently, with $1 \leq j \leq m$, and in this way obtain decisions x_j, with $1 \leq j \leq m$, such that

$$f_j(x_j) = \sup_{x \in X} f_j(x), \quad 1 \leq j \leq m \tag{2}$$

it may well happen that quite a few of the resulting decisions $x_1, \ldots, x_m$ are *different* due to the fact that the objectives (1) are conflicting, thus they cannot be maximized *simultaneously*.

For a better understanding of the above difficulty we shall proceed to a further refinement of the basic model (1). There are several customary ways this can be done and we shall present two of the most well known ones.

A possible refinement of the model (1) can be achieved by introducing a *preference structure* on the set X of feasible decisions. Under sufficiently general conditions, [B. Roy], this can be done by specifying *four binary relations* on X.

$$I, P, Q, R \subset X \times X \tag{3}$$

called respectively, *indifference, strict preference, large preference and incomparability*, which can for instance, have the following interpretation. Given the feasible decisions, $x, x' \in X$, let us denote

$$C(x,x') = \{1 < j < m \mid f_j(x) = f_j(x')\} \tag{4}$$

$$D(x,x') = \{1 < j < m \mid f_j(x) \neq f_j(x')\} \tag{5}$$

and call them the *concordance* respectively *discordance* sets of objectives relative to x and x'. Then we can say that we are *indifferent* between x and x' and write $(x,x') \in I$ or xIx', if

$$|f_j(x) - f_j(x')| \text{ negligible, for } j \in D(x,x') \tag{6}$$

Further, we can say that x' is *strictly preferred* to x and write $(x,x') \in P$, or xPx', if for $j \in D(x,x')$ we have

$$\begin{aligned} &\text{either } |f_j(x') - f_j(x)| \text{ negligible} \\ &\text{or } f_j(x') - f_j(x) \text{ large} \end{aligned} \tag{7}$$

Then, we can say that x' is *largely preferred* to x and write $(x,x') \in Q$ or xQx', if for $j \in D(x,x')$ we have

$$\begin{aligned} &\text{either } |f_j(x) - f_j(x')| \text{ negligible} \\ &\text{or } f_j(x') - f_j(x) > 0 \text{ but neither negligible nor large} \end{aligned} \tag{8}$$

Finally, we can say that x and x' are *incomparable* and write $(x,x') \in R$ or xRx', if

$$\begin{aligned} & f_j(x') - f_j(x) > 0 \quad \text{for some} \quad j \in D(x,x') \\ & \text{and in the same time} \\ & f_j(x) - f_j(x') > 0 \quad \text{for some} \quad j \in D(x,x') \end{aligned} \tag{9}$$

and the respective positive quantities are not negligible. It follows that I is reflexive and symmetric, P and Q are irreflexive and antisymmetric while R is irreflexive and symmetric. Moreover, under very general conditions, it can be assumed that for every given $x,x' \in X$ one and only one of the following *six alternatives* holds

$$xIx', \; xPx', \; x'Px, \; xQx', \; x'Qx, \; xRx' \tag{10}$$

It is particularly *important* to point out that within the above refinement (3 - 10) of the basic model (1) the following *two* essential features will in general accompany the presence of conflicting objectives

P and/or Q may be *intransitive* (11)

the *incomparability* relation R may be present (12)

Indeed, whenever (11) and (12) do not hold, i.e., P and Q are transitive and no feasible decisions $x,x' \in X$ are incomparable, it follows easily that we *cannot* have conflicting objectives, therefore for all practical purposes, we are in the presence of a *classical optimization* problem. It is for this reason (among others) that the traditional method of *priorities* may present disadvantages, as it *precludes* the possibility of both *intransitivity and incomparability* within the preference structure to which it may lead on the set of feasible decisions X. Indeed, according to this method, first we have to order the $m \geq 2$ objectives in (1) according to their importance or priority. For simplicity, let us assume that this ordering has already been made and it has resulted in the order in (1). Then, we can define on X a preference structure (3) as follows. The indifference relation xIx' means

$$|f_j(x') - f_j(x)| \text{ negligible, for } 1 \leq j \leq m \tag{13}$$

Further, the strict or large preference relation $x(P \cup Q)x'$ means that for a certain $1 \leq k \leq m$ we have

$$\begin{aligned} &|f_j(x') - f_j(x)| \text{ negligible, for } 1 \leq j < k \\ &\text{and in the same time} \\ &f_k(x') - f_k(x) > 0 \text{ and not negligible} \end{aligned} \tag{14}$$

Then it is easy to see that the six alternatives in (10) reduce to the following three

$$xIx', \quad x(P \cup Q)x', \quad x'(P \cup Q)x \tag{15}$$

in other words (12) will not hold. It also follows that $P \cup Q$ is transitive and (11) is also not valid. In conclusion, the main disadvantage of the method of priorities is that it leads to a *too early and simple* elimination of conflict when from the very beginning it orders the objectives according to their so called importance or priority.

Another possible refinement of the model (1) can be obtained by the introduction of a *utility function*

$$U : Y \to R^1 \tag{16}$$

where we denoted

$$Y = \{(f_1(x),\dots,f_m(x)) \mid x \in X\} \subset R^m \tag{17}$$

This refinement contains as a particular case the traditional method of *weighted objectives* which is obtained when the utility function (16) is assumed to have the following *linear* form

$$U(f_1(x),\dots,f_m(x)) = \sum_{1 \leq j \leq m} w_j f_j(x), \; x \in X, \tag{18}$$

where

$$w_1,\dots,w_m \geq 0, \; w_1 + \dots + w_m = 1 \tag{18.1}$$

are the respective weights. As is well known, the utility function approach (16, 17) presents *two major disadvantages*. First, it often proves unrealistic to assume that the DM is ready to express his *global* utility function (13), i.e., to specify its values on the entire set Y. Then, even in the case where this could be done, the result is a preference structure on X which, similarly to the case of the method of priorities, may prove to be too simple since it *cannot* accommodate intransitivity or incomparability. Indeed, in this case the preference structure on X can be defined as follows. The indifference relation xIx' means

$$|V(x') - V(x)| \text{ negligible} \tag{19}$$

where we defined

$$V : X \to R^1 \tag{20}$$

by

$$V(x) = U(f_1(x),\dots,f_m(x)),\ x \in X \tag{20.1}$$

Further, the strict or large preference relation $x(P \cup Q)x'$ means

$$V(x') - V(x) > 0 \text{ and not negligible} \tag{21}$$

Then it is easy to see again that the six alternatives in (10) reduce to the three in (15), moreover (11) and (12) will not hold.

As mentioned at the beginning, Pareto suggested one of the first ways for avoiding the above disadvantages of the two traditional methods of dealing with conflicting objectives. He suggested as solution concept the notion of *nondominated decision* which can be defined as follows. Suppose given the preference structure (3) on the set X of feasible decisions. Then a feasible decision $x \in X$ is called *nondominated*, if for any other feasible decision $x' \in X$ we have

$$x(P \cup Q)x' \Rightarrow xIx' \tag{22}$$

Let us denote by X^P the subset of nondominated feasible decisions in X.

As is well known, one of the major disadvantages of the above solution concept is that the set X^P of nondominated decisions may prove to be *too large* to be acceptable as the set of optimal decisions. In order to make this more clear, we shall indicate *two* possible meanings of being 'too large', both of which can manifest themselves when dealing with specific instances of the nondominated set X^P in practical applications. Indeed, let us suppose the framework of (1, 3 - 10, 17) and similarly to (22), let us define

$$Y^P = \left\{ y \in Y \;\middle|\; \begin{array}{l} \forall \; y' \in Y : \\ y \leqslant y' \Rightarrow y' = y \end{array} \right\} \tag{23}$$

where for $y = (y_1, \ldots, y_m)$, $y' = (y'_1, \ldots, y'_m) \in R^m$ we denote $y \leqslant y'$, if $y_j \leqslant y'_j$, for $1 \leqslant j \leqslant m$. Further, let us define the mapping

$$f : X \to Y \ , \ f(x) = (f_1(x), \ldots, f_m(x)), \ x \in X \tag{24}$$

Then it is obvious that

$$Y^P \subset f(X^P) \subset \text{neighbourhood of } \ Y^P \tag{25}$$

and this relation can imply a rather large X^P.

Indeed, connected with the left inclusion, in (25) we notice that in view of (23) it follows that Y^P is a *m - 1 dimensional* manifold in R^m. Thus for a number of objectives

$$m \approx 10 \tag{26}$$

the set Y^P has an inconveniently *high dimension*. In addition, the *reduction of dimension* from m to $m - 1$ obtained when going from Y to Y^P is *marginal*. However, a difficulty of no less importance arises in connection with the right inclusion in (25). Indeed, in case of a larger number m of objectives, a neighbourhood of Y^P can occupy a *significant part* of the whole volume of Y. For the sake of simplicity, let us assume given $\rho > 0$ and

$$Y = \left\{ y = (y_1, \ldots, y_m) \in R^m \;\middle|\; \begin{array}{l} *) \ y_1, \ldots, y_m > 0 \\ **) \ y_1 + \ldots + y_m < \rho \end{array} \right\} \tag{27}$$

then obviously

$$Y^P = \{y = (y_1, \ldots, y_m) \in Y \mid y_1 + \ldots + y_m = \rho\} \tag{28}$$

while for given $0 < \varepsilon < \rho$ we have

$$Y^P(\varepsilon) = \{y = (y_1,\dots,y_m) \in Y \mid \rho - \varepsilon \leq y_1 + \dots + y_m \leq \rho\} \tag{29}$$

where we denoted by $Y^P(\varepsilon)$ the ε-neighbourhood of Y^P in Y. In this case an easy calculation yields

$$\frac{\text{vol } Y^P(\varepsilon)}{\text{vol } Y} = 1 - \left(1 - \frac{\varepsilon}{\rho}\right)^m \tag{30}$$

If we take for instance

$$\varepsilon = \rho/m \ , \ m = 20 \tag{31}$$

then (30) yields

$$\text{vol } Y^P(\varepsilon) \approx .63 \text{ vol } Y \tag{32}$$

therefore a neighbourhood of Y^P with a thickness of 5% of the height of Y will contain about 63% of the volume of Y.

It can be said that to a large extent the above mentioned difficulties connected with the traditional methods of priorities, weighted objectives, utility functions or nondominated solutions were behind later attempts to develop more efficient and realistic ways for dealing with decisions in a context of conflicting objectives.

The aim of this study is to present on the basis of the author's recent research, [Rosinger, a,b] and [Mond and Rosinger a,b,c] a direction of development for efficient and realistic aids for decision making with conflicting objectives.

1. A PRIORI VERSUS ALGORITHMIC SOLUTION CONCEPTS

One of the important consequences which follow from the previous Section is the crucial role which the *solution concepts* employed play in decisions with conflicting objectives. This situation is characteristic for such decision processes, since in the case of one single objective or several nonconflicting objectives, there is an *unique*, classical solution concept which can be employed, the only problem left being to devise suitable methods for reaching or at least approaching the optimal solution.

As can be seen arriving at suitable solution concepts is not a trivial task. Indeed, we have seen that the solution concepts resulting from the traditional methods of priorities, weighted objectives or more generally, utility functions, suffer, among other things, from the fact that they lead to a *too early and simple* solution of the conflict. On the other hand, the solution concept of Pareto leaves *too much* of the conflict unsettled. It follows that suitable solution concepts should somehow be placed between these two kinds of extremes.

As an *important, general remark* we should mention that the *choice* of a suitable solution concept is in *itself* a decision with conflicting objectives. For the sake of clarity we shall call them *metadecisions* and *metaobjectives* respectively. In this respect, *two* of the most important metaobjectives which a solution concept should satisfy are

a) ease in finding the optimal solutions

b) fidelity in modelling the conflict between the initial objectives.

These two metaobjectives are obviously shared both by those who devise decision aids and those, for instance DM-s, who are going to use them.

Several other relevant metaobjectives have been suggested in the literature and have been used in order to test various aids for decisions with conflicting objectives, as mentioned for instance in [J. Wallenius], [Rosinger a,b].

In these terms, the traditional methods of priorities, weighted objectives or utility functions perform very well in so far as the metaobjective a) is concerned, but perform rather poorly with regard to the metaobjective b). On the other hand, the Pareto solution concept exhibits exactly the opposite behaviour.

The process of devising solution concepts which are satisfactory from the point of view of both of the above metaobjectives a) and b) is significantly helped by a study of some of the *general requirements* a DM would prefer to have satisfied when using a decision aid. In this respect, on the basis of an accumulated extensive experience on the ways DM-s regard various aids for decisions with conflicting objectives it became apparent, [H. Wallenius], that there exists a preference for decition aids based on solution concepts which do *not preclude* the *participation* of the DM in the particular decision processes. In other words, there exists an aversion towards 'a priori solution concepts' for which the corresponding decision aids behave like a 'black box', i.e., deliver certain optimal decisions without giving the DM any insight along the way as to how these optimal decisions were chosen. Obviously, this aversion is a direct consequence of the lack of confidence the DM has in optimal decisions which were obtained in such a way. It follows that a particularly important metaobjective of a DM is

c) the confidence in the solutions obtained.

It should be pointed out that the desire of a DM to participate more intimately in a particular decisions process is to a large extent a fortunate situation. Indeed, it will help avoid the situation where prior to the particular decision process, the DM would be asked to supply *all* the relevant information, a demand which could prove to be unrealistic, as seen for instance in the previous Section. In other words, the DM's desire to participate in the decision process gives the opportunity for *several* inquiries about his *local preferences*. However, it is obvious that in connection with his participation, the DM has the following two somewhat conflicting *metaobjectives*

d) ease of participation

e) efficiency of participation.

During the last decade, a wide variety of so called *interactive* aids for decisions with conflicting objectives have been developed with the explicit or implicit aim of possibly reconciling the above metaobjectives a) - e), as well as other ones, [H. Wallenius].

In Sections 2 - 4 we shall present a succession of *three* aids for decisions with conflicting objectives, [Rosinger a,b], [Mond & Rosinger, a,b,c], each of them seriously improving on metaobjective d), without at the same time losing too much on metaobjectives a) - c) and e). This is done by successively diminishing the degree of *on-line* participation of the DM within the respective interactive processes. Indeed, while the first aid requires the maximum on-line participation in the sense that the DM *has* to provide a certain information on his local preference at *each* iteration step, the last aid does *not* necessarily require on-line participation as the DM could, in fact, let the solution algorithm operate alone, if he wished to do so.

It should be pointed out that the idea of devising aids for decisions with conflicting objectives based on a conscious *metadecision* process involving possibly conflicting *metaobjectives* has emerged recently. The present study offers one possible approach which is complementary to the approach in [Gershon & Duckstein]. In fact we try rather to *devise additional* decision aids which may be *metanondominated*, i.e., nondominated in the metadecision process over the given metaobjectives while the approach in [Gershon & Duckstein] only *chooses* among existing decision aids, by building a particular metadecision process for the purpose.

We observe that metaobjectives a) and b) have a rather 'objective' character in the sense that they may arise from the point of view of *everybody* who is in some way or other involved in the decision process. In particular, they are valid both for those who devise decision aids and those who will use them, i.e., the DM's. On the other hand, metaobjectives c) - e) have a more 'subjective' character, since they arise mainly from the point of view of the DM. In this connection it is important to note the *priority* of metaobjective e) which becomes the driving force behind the tendency to go from *a priori solution concepts* which preclude the DM's participation towards various *algorithmic solution concepts*, such as those base on *interactions* of the DM. The essential feature of these latter solution concepts is that they do *not* give a *definition* of the optimal solution and only give a *method* for obtaining it.

2. A FULLY ON-LINE INTERACTIVE DECISION AID

In [Rosinger a,b] we presented an improvement of the well known interactive decision aid proposed in [Geoffrion et al.]. In this Section we shall shortly recapitulate the main features of this improved method.

The model of the decision problem is that in (1,16,17) and our problem is that the DM is *not* expected to state his *global* utility function (16). Therefore the decision problem cannot be reduced to the following classical optimization problem

$$\sup_{x \in X} U(f_1(x),\ldots,f_m(x)) = ? \qquad (33)$$

However, we shall expect that the DM is ready to participate in *each* step of an iterative process and provide certain *local* information on his utility function (16). In this connection, a *first* important fact remarked in [Geoffrion et al.] is that in case we use a gradient method for solving (33), such as for instance the Frank-Wolfe steepest ascent algorithm, we do *not* need the full knowledge of the global utility function (16). Indeed, the mentioned algorithm constructs a sequence

$x_0, x_1, \ldots, x_k, x_{k+1}, \ldots \in X$ which converges to an optimal solution $x_* \in X$ and it proceeds according to the following iterative steps.

$$\text{Choose } x_0 \in X \text{ arbitrarily} \tag{34}$$

If the iterations $x_0, \ldots, x_k \in X$ have already been obtained, then as the next step in the iteration process construct $x_{k+1} \in X$ according to the following three substeps. First, find $y_k \in X$ a solution of

$$\sup_{y \in X} \sum_{1 \leq j \leq m} (\partial/\partial f_j) U(f_1(x_k), \ldots, f_m(x_k)) \cdot \nabla f_j(x_k) \cdot y \tag{35}$$

Then, find $t_k \in [0,1]$ a solution of

$$\sup_{t \in [0,1]} U(f_1(x_k + t(y_k - x_k)), \ldots, f_m(x_k + t(y_k - x_k))) \tag{36}$$

Finally, take

$$x_{k+1} = x_k + t_k(y_k - x_k) \tag{37}$$

A *second* important fact remarked in [Geoffrion et al.] is that it is a rather *easy* task for the DM to provide the information on his utility function (16) needed in substep (36), i.e., to choose $t_k \in [0,1]$. In view of this, it will be assued that at each iteration k the DM can give the value of t_k. However, the same thing *cannot* be said about the information on the utility function (16) needed in substep (35). In this case the information needed is the *direction* of the m-dimensional gradient vector

$$a_k = (a_{k1}, \dots, a_{km}) \in R^m \tag{38}$$

where

$$a_{kj} = (\partial/\partial f_j) U(f_1(x_k), \dots, f_m(x_k)), \quad \text{for} \quad 1 \leq j \leq m \tag{38.1}$$

A *third* important fact remarked in [Geoffrion et al.] is that as far as we are only interested in the direction of the m-dimensional vector (38), it will be sufficient if the DM gives the m - 1 dimensional vector

$$b_k = (b_{k1}, \dots, b_{km-1}) \in R^m \tag{39}$$

where

$$b_{kj} = a_{kj+1}/a_{k1} \quad , \quad \text{for} \quad 1 \leq j \leq m-1 \tag{39.1}$$

assuming for instance that

$$a_{k1} > 0 \tag{39.2}$$

The point in asking the DM to give the vector b_k instead of the vector a_k is not the reduction in dimension from m to $m-1$ which in many practical situations when m is as in(26) is marginal, but the fact that the components of b_k are *relative marginal utilities* and it is expected that the DM will find it much *easier* to provide that latter type of information.

However, in many practical situations even if m is not large, it may prove to be difficult to compare the marginal utilities required in (39.1), where *all* marginal utilities have to be related to *one* and the same given marginal utility. Indeed, it may happen that the comparison of marginal utilities does not quite make sense for certain pairs or even groups of objectives.

The method in [Rosinger a,b] was devised in order to overcome the above mentioned difficulty by offering the DM a *large framework* within which he can not only *give* his answers needed in the substep (38) but can also *choose* the questions he is to answer. Moreover, the DM will be able to compare the marginal utilities of certain *groups* of objectives only. What proves to be particularly convenient, in case certain objectives appear in several such groups, is that the DM can give answers which are *contracitctory* to a certain extent. This depends only on the *robustness* of the convergence the underlying Frank-Wolfe steepest ascent algorithm. This framework which contains as a rather severe particular case the inquiry in (39) needed in the method of [Geoffrion et al.] will be set up next. Since the problem of answering (38) is the same at any iteration k, that index will no longer be mentioned.

Then we notice that from a strictly mathematical point of view the problem of answering (38) becomes the problem of *constructing* a framework for inquiries aimed at *identifying* a point on the surface of the unit ball in R^m

$$a = (a_1,\dots,a_m) \in R^m \ , \quad \|a\| = 1 \ , \tag{40}$$

where $\| \ \|$ is any given norm on R^m.

A convenient and wide framework for the inquiries can be obtained as follows. First, the DM selects a number $p \geq 1$ of subsets of at least two elements

$$J_1,\dots,J_p \subset \{1,\dots,m\} \tag{41}$$

representing *groups of objectives* such that he is ready to compare the marginal utilities of any two objectives within each of the above groups. Then the DM gives p vectors

$$d_i = (d_{ij} \mid j \in J_i) \ , \quad \text{for} \quad 1 \leq i \leq p, \tag{42}$$

trying to fulfil as much as possible the following p conditions

$$d_i \ \text{ has the direction of } \ (a_j \mid j \in J_i), \quad \text{for} \quad 1 \leq i \leq p \tag{43}$$

Two examples are presented in order to clarify the meaning of the inquiry in (41 - 43).

Example 1

Let us assume that the DM takes $p = 1$ and $J = \{1,\dots,m\}$. Then he has to give a vector $d_1 = (d_{11},\dots,d_{1m})$ which has as far as possible the direction of $a = (a_1,\dots,a_m)$. Obsiously, this is the *hardest* way to answer (38).

Example 2

Let us assume that the DM takes $p = m - 1$ and $J_i = \{1, i+1\}$, for $1 \leqslant i \leqslant p$. Then he has to give the vectors $d_i = (d_{i1}, d_{ii+1})$, for $1 \leqslant i \leqslant p$, having as far as possible the respective directions (a_1, a_{i+1}), for $1 \leqslant i \leqslant p$. It is obvious that in case $a_1 > 0$, the above inquiry is the same with (39) which is requested in [Geoffrion et al.], thus by requiring the statement of a $m-1$ dimensional vector it is only marginally easier than that in Example 1.

In the sequel it will be convenient to reformulate the inquiry in (41 - 43) with the help of matrices. In this respect we notice that the subsets in (41) can be put in a one-to-one correspondence with a matrix

$$P = (P_{ij} \mid 1 \leqslant i \leqslant p,\ 1 \leqslant j \leqslant m) \tag{44}$$

acoording to

$$P_{ij} = \begin{cases} 1 & \text{if } j \in J_i \\ 0 & \text{if } j \notin J_i \end{cases} \tag{44.1}$$

Any matrix (44) is called in *inquiry pattern* if

$$\forall \quad 1 \leqslant i \leqslant p : \exists \quad 1 \leqslant j < \ell \leqslant m : P_{ij} = P_{i\ell} = 1 \tag{45}$$

$$\forall \quad 1 \leqslant j \leqslant m : \exists 1 \leqslant i \leqslant p : P_{ij} = 1 \tag{46}$$

Condition (45) means that each group of objectives J_i contains at least two objectives, while condition (46) means that each objective j belongs to at least one group of objectives J_i.

Now, the vectors (42) can be written in the form of a matrix with real entries

$$D = (D_{ij} \mid 1 \leqslant i \leqslant p, \ 1 \leqslant j \leqslant m) \tag{47}$$

where

$$D_{ij} = \begin{vmatrix} d_{ij} & \text{if} \quad j \in J_i \\ 0 & \text{if} \quad j \notin J_i \end{vmatrix} \tag{47.1}$$

The characteristic property of these matrices is that for $1 \leqslant i \leqslant p$, $1 \leqslant j \leqslant m$ we have

$$P_{ij} = 0 \Rightarrow D_{ij} = 0 \tag{48}$$

or equivalently

$$P_{ij} D_{ij} = D_{ij} \tag{49}$$

We shall call any such matrix D a *P-answer*.

With the above definitions, the inquiry in (41 - 43) can be reformulated as follows. The DM chooses an inquiry pattern P and then gives a P-answer D so that for suitable $\lambda_1,\ldots,\lambda_p > 0$, the relation

$$P \begin{pmatrix} a_1 & & 0 \\ & \ddots & \\ 0 & & a_m \end{pmatrix} = \begin{pmatrix} \lambda_1 & & 0 \\ & \ddots & \\ 0 & & \lambda_p \end{pmatrix} D \qquad (50)$$

is fulfilled as nearly as possible.

We can reformulate the above Examples 1 and 2 in terms of (44 - 50). In the first example we shall have

$$p = 1 \quad \text{and} \quad P_{ij} = 1 \,, \quad \text{for} \quad 1 \leqslant j \leqslant m \qquad (51)$$

Therefore, given a P-answer D, there exists an *unique* vector (40) which for suitable $\lambda_1,\ldots,\lambda_p > 0$ satisfies (50), if and only if

$$D \neq 0 \qquad (52)$$

In the second example we shall have

$$p = m-1 \quad \text{and} \quad P_{ij} = \begin{cases} 1 & \text{if} \quad j = 1 \quad \text{or} \quad j = i+1 \\ 0 & \text{otherwise} \end{cases} \qquad (53)$$

It follows easily that given a P-answer D, there exists an *unique* vector

(40) which for suitable $\lambda_1,\ldots,\lambda_p > 0$ satisfies (50), if and only if

$$D_{11}D_{i1} > 0\ , \quad \text{for} \quad 1 \leqslant i \leqslant p \tag{54}$$

Going back to the general inquiry (44 - 50), it is particularly *important* to note that in the practical application of the interactive process, the relation (50) is actually used in a *converse* way. Indeed, as the vector (40) is not known, the DM chooses P and states D and then the decision aid has to find a vector (40) which for suitable $\lambda_1,\ldots,\lambda_p > 0$ will satisfy (50) as nearly as possible. It should be noted that in these terms the equation (50) does *not* necessarily have an *exact solution* (40) since it contains the positivity conditions on λ_i and moreover, it may be overdetermined as the number $P_{11} + \ldots + P_{pm}$ of equations may be larger than the number $m+p$ of unknowns a_j and λ_i. For this reason, equation (50) will be replaced by the problem of *minimizing* in a certain norm the matrix

$$P \begin{pmatrix} a_1 & & 0 \\ & \ddots & \\ 0 & & a_m \end{pmatrix} - \begin{pmatrix} \lambda_1 & & 0 \\ & \ddots & \\ 0 & & \lambda_p \end{pmatrix} D \tag{55}$$

over all vectors (40) and $\lambda_1,\ldots,\lambda_p > 0$ and with given P and D.

As a *conclusion* we can mention the following. As shown in tests with professional DM's in situations near to their sphere of competence as well as in real applications, the above fully interactive decision aid performs well where the meta-objectives (a), (b), (c) and (e) are concerned. The two features of the aid particularly enjoyed by DM's were the *freedom* offered in setting up the way the inquiries are conducted and the possibility of giving answers which to a certain extent are *contradictory*, neither of these features being offered by the aid in [Geoffrion et al.]. A difficulty which arises in the practical application of the above aid and which could not happen with the aid in [Geoffrion et al.] comes from the freedom the DM has in setting up the inquiry. Whenever the iterations start with a feasible decision $x_0 \in X$ which is somewhat familiar to the DM, it will be quite easy for him to choose the initial inquiry pattern p_0. However, after a number of iterations k, this inquiry pattern need no longer be proper. The *difficulty* which seems to arise is the *recognition* of such a moment k in the succession of iterations and then, the proper *choice* of a new inquiry pattern p_k.

A decision aid aimed at helping in this situation was suggested in [Mond and Rosinger a,b]. Its basic ideal will be presented next.

3. A DECISION AID WITH EASIER ON-LINE INTERACTION

The fully interactive decision aid presented in Section 2 assumed the existence of an *utility function* (16), even if the DM was only assumed to have an implicit and local knowledge of it, which proved to be sufficient to set up a Frank-Wolfe type iterative-interactive process converging to an optimal decision.

In case we want to avoid the relatively lengthy on-line participation of the DM and replace it by a shorter one, a further particularization in (18) of the model in (1,16) may be useful. Indeed, in principle, the classical method of weighted objectives does in fact require *one single* iteration, i.e., the DM is requested only to provide the weights (18.1).

However, in order to avoid the *principal* deficiencies of this classical method, among others the lack of *intransitivity* and *incomparability*, we could combine it with a *relatively short* interactive process. The way such an interactive process may be set up can be suggested by the *practical* deficiencies of the mentioned method; deficiencies which may thus be diminished. Indeed, it is well-known that the main difficulty the DM encounters when using the method of weighted objectives is that he is required to state *directly* the weight $w_j > 0$ for each individual objective $1 \leq j \leq m$ and this statement does *not* allow for a sufficient consideration of the possible *conflict* between various objectives, except for interactions through the *normalization* condition

$$w_1 + \ldots + w_m = 1 \tag{56}$$

or through a *relatively small* number of pair-wise comparisons

$$w_j/w_\ell\,, \quad \text{for } \textit{certain } 1 \leq j < \ell \leq m. \tag{57}$$

In other words, the DM is not given a sufficient chance to satisfy meta-objective (b).

The usual way of avoiding this practical difficulty is to give the

DM the opportunity to state the weights (18.1) in an *indirect* way. Several such methods, for instance based on priorities in hierarchical structures [Saaty a], or hierarchical additive weighting, hierarchical tradeoff, etc., have been presented in the literature, as surveyed recently in [Hwang and Yoon]. These methods are usually based on the introduction of additional structure in the space of decisions X through certain preliminary inquiries the DM has to answer.

The aim of the method in [Mond and Rosinger a,b], presented shortly in the sequel, is to offer the DM a *large framework* - similar to the one in Section 2 - which gives him a *significant freedom* in both stating the weights (18.1) in an *indirect* way and expressing his feelings about various possible *interdependencies* between two or more objectives which may be conflicting. Additional advantages of this method are the following: the DM does *not* have to *structure* the space of decisions X, and if the DM participates in at least *two* interactions, the nonrealistic elimination of intransivity and incomparability specific to the classical method of weighted objectives can be overcome to a large extent.

The basic idea of the method is to ask the DM at a given interation such questions on the weights (18.1) as can offer him as wide a choice as possible to give *indirect*, somewhat *contradictory* answers which, however, allow for the manifestation of certain sophisticated *interdependendencies* between the possibly conflicting objectives, beyond the usual interdependencies which can be expressed in (56, 57), for instance. In this connection it should be mentioned that the method in [Saaty a] for instance, is based on the requirement of stating *all* the pair-wise comparisons

$$w_j/w_\ell \text{ , for } all \quad 1 \leq j < \ell \leq m. \tag{58}$$

This has the following *two* disadvantages. First, in case the number m of objectives is *large*, the information required by (58) becomes too *demanding* on the DM. Second, the mentioned method does *not* allow for *other* comparisons, except for the pair-wise ones in (58). In connection with the elimination of the first disadvantage a recent empirical method was proposed in [Saaty b].

Going back to the method in [Mond and Rosinger a,b], the above-mentioned goals are achieved thus. The inquiry facing the DM when he is required to state the weights (18.1) is embedded into a *wider family of inquiries* in which the DM is free both to *choose* the questions and *give* the answers, similar to the situation in Section 2. Indeed, at any given iteration, the DM is requested to select a number $p \geq 1$ of *groups of objectives*

$$J_1, \ldots, J_p \subset \{1, \ldots, m\}. \tag{59}$$

This is supposed to be done in such a way that the DM is then ready to state *joint weights*

$$W_1, \ldots, W_p \geq 0 \tag{60}$$

which have the meaning that his *cardinal preference* for the totality of objectives j in J_i is W_i, for $1 \leq i \leq p$. It is important to note that in case of the information in (60), we shall *not* need a normalization condition.

As an illustration we mention that the classical weights in (18) correspond to the particular case of (59, 60) when

$$p = m,\ J_1 = \{1\},\ldots,J_m = \{m\},\ W_1 = w_1,\ldots,W_p = w_p. \tag{61}$$

Obviously, the case in (58) used in [Saaty a] is also a particularization of (59, 60).

The only conditions on (59, 60) which we require are

$$\phi \neq J_i \subseteq \{1,\ldots,m\}\ ,\quad \text{for}\quad 1 \leq i \leq p \tag{62}$$

$$J_1 \cup \ldots \cup J_p = \{1,\ldots,m\} \tag{63}$$

$$J_i \subseteq J_h \Rightarrow W_i \leq W_h\ ,\quad \text{for}\quad 1 \leq i,h \leq p \tag{64}$$

which are obviously satisfied in the classical case (61). Obviously, (63) is the same with (45). Further, in view of (62, 63) it follows that we may assume

$$2 \leq p \leq 2^m - 1 \tag{65}$$

however, in order to avoid subjecting the DM to excessive demenads, we expect that

$$p = O(m) \tag{66}$$

as happens for instance in the classical case (61).

At this stage it is convenient to introduce two definitions. The information in (59, 60) can be written in the matrix form

$$W = \begin{matrix} J_1 & \cdots\cdots & J_p \\ W_1 & \cdots\cdots & W_p \end{matrix} \tag{67}$$

A matrix (67) which satisfies (62 - 64) will be called a *set of interdependent weights*. Further, the particular case of (67) when

$$p = m \ , \ J_1 = \{1\},\ldots,J_m = \{m\} \tag{68}$$

yields

$$w = \begin{matrix} 1 & \cdots\cdots & m \\ w_1 & \cdots\cdots & w_m \end{matrix} \tag{69}$$

which will be called *a set of simple weights*.

With the above definitions, we can now formulate the mathematical problem: the DM is asked to give a *set of interdependent weights* W and then the decision aid has to find a *set of simple weights* w which in a certain sense, to be specified next, is the *best approximation* to W.

We note that the information W given by the DM could possibly be used in various other ways in order to set up an utility function (16). However, here we shall consider only the above way which leads to the particular form of the utility function in (18).

In connection with the mentioned best approximation of the given W by an arbitrary, unknown w it is obvious that the *ideal* situation would obtain when

$$W_i = \sum_{j \in J_i} w_j \ , \quad \text{for} \quad 1 \leq i \leq p \tag{70}$$

for a suitable choice of the set of simple weights w. However, the system of linear equations (70) may happen to be overdetermined or underdetermined in the unknowns $w_1,\ldots,w_m$. In view of this, the approximation of W by w can be done by minimizing in w the *quadratic function*

$$E_{W,\alpha}(w) = \sum_{1 \leq i \leq p} \alpha_i \Big(\sum_{j \in J_i} w_j - W_i \Big)^2 \tag{71}$$

or

$$F_{W,\alpha}(W) = \sum_{1 \leq i \leq p} \frac{\alpha_i}{W_i} \Big(\sum_{j \in J_i} w_j - W_i \Big)^2 \tag{72}$$

where

$$\alpha = (\alpha_1,\ldots,\alpha_p) \ , \ \alpha_1,\ldots,\alpha_p > 0 \tag{73}$$

are either given by the DM or for instance, taken all equal to 1. As is known, the minimization of the function in (72) could have the advantage of diminishing undesirable consequences of the presence of some excessively large joint weights W_i.

The need for *further iterations*, based on the DM's interaction, may arise for the following two reasons: the solution w obtained from the version used of the above minimization problems is *not unique*, or even if the solution w is unique, the DM is *not* satisfied with the resulting decision, obtained from the maximization of the utility function defined by w. In these cases the DM can interact in either of the following ways.

1. In case of nonunique w, the range of different solutions (which will now be much more limited in scope than the original range of possible weights) can be presented to the DM who may select that solution which seems most appropriate to him.

2. The DM can be asked to provide *additional* groups of attributes $J_{p+1}, J_{p+2}, \ldots,$ and corresponding additional joint weights $W_{p+1}, W_{p+2}, \ldots$ to augment those already provided in (67). This should not prove too difficult since there are, all together, $2^m - 1$ possible groups. It may also be possible to indicate to the DM, on the basis of the solutions already obtained, some of the groups of attributes for which additional weights would be most helpful. Note that in this case, if unequal weights α in (73) had been provided, a new set of such α will now be needed. It is also possible to ask the DM to *replace* some of the initially stated groups of

attributes J_i and corresponding joint weights W_i, with $1 \leq i \leq p$, by one or more other groups of attributes and their joint weights. Finally, it is also possible to ask the DM to *cancel* some of the initially stated groups of attributes and corresponding joint weights. The above three ways of adding, replacing or cancelling certain information can prove to be particularly useful for a *sensitivity or stability analysis* of a given solution w.

If it is decided not to ask the DM for additional information, or if he declines to provide it, it is probably best since the solution set is convex, to use an average of the extreme point solutions already obtained.

It should be noted however that the nonuniqueness of the solution w does not necessarily represent a disadvantage if considered in a larger context. Indeed, it may well happen that the DM will have an interest in *intervals* rather than single values over which the simple weights $w_1,\dots,w_m$ can range. The reason for this is that in such a situation the respective solutions $x \in X$ resulting from the maximization of the corresponding utility function (18) can give a better view of the multiple attribute decisionmaking situation, or simply can provide a new and reduced subset $X_1 \subset X$ for starting a new iteration.

We illustrate the procedure by the following simple example. Suppose there are five objectives to be minimized and the DM initially provides the following attributes and corresponding weights.

$W(\{1,2\}) = 3$, $W(\{1\}) = 1$, $W(\{1,2,3\}) = 4$, $W(\{3\}) = 2$, $W(\{4,5\}) = 8$

Assume that no further weights α are provided so that we take all $\alpha_i = 1$.

The solution of the quadratic programming problem is

$$\lambda(1 \ , \ 5/3, \ 5/3 \ , \ 8 \ , \ 0) + (1-\lambda)(1 \ , \ 5/3 \ , \ 5/3 \ , \ 0 \ , \ 8) \ ,$$

for all λ, $0 \leqslant \lambda \leqslant 1$.

This means that the simple weights w_1, w_2, w_3 are uniquely determined but that the simple weights w_4 and w_5 must satisfy

$$w_4 + w_5 = 8 \ , \ w_4 \geqslant 0 \ , \ w_5 \geqslant 0. \tag{74}$$

If it is decided not to ask the DM for further information, or if he declines to provide it, it is suggested that one takes $w_4 = w_5 = 4$. On the other hand equation (74) can be presented to the DM who may choose other values for w_4 and w_5 satisfying (74). Alternatively the DM can provide additional groups of attributes and corresponding weights. It should be clear from (74) that a group of attributes involving w_4 or w_5, but not both, would be most helpful here.

Suppose now the DM provides the additional weight

$$W(\{2,5\}) = 7.$$

The quadratic programming problem now has the unique solution

$$(1 \ , \ 5/3 \ , \ 5/3 \ , \ 8/3 \ , \ 16/3).$$

4. A DECISION AID WITH ON-LINE, OFF-LINE INTERACTION

In case the on-line demands during the interactive decision aids presented in Section 2 and 3 prove to be inconvenient, the DM can opt for the following decision aid based on *clustering techniques*, which offers the possiblity of a *free choice* of the *degree of interaction*, ranging between full on-line on one side and off-line on the other side.

The idea of using a clustering technique (CT) as an aid for decisions with conflicting objectives involving one single decision-maker was first suggested in [Rivett]. At least in principle, the suitability of this idea is obvious as the solution of such a decision problem is in fact a ranking of a set of feasible decisions according to their desirability. As the respective desirability is judged in view of several possibly conflicting objectives, the set of feasible decisions is constituted into a complex system of multiple and simultaneous rankings in which the most desirable decisions can be seen as a *special cluster*. An obvious advantage of this approach is in the fact that in general, a CT requires *less information* than a usual ranking method which would for instance, explicitly or implicitely assume the global or local knowledge of an utility function (16). This advantage is clearly illustrated in [Rivett], where given the finite set X of feasible decisions, the only information required is the indifference set $I \subset X \times X$, i.e., the set pairs $(x,y) \in X \times X$ such that the decision-maker DM is indifferent between the outcomes of decisions x and y. However, the approach in [Rivett] does not make full use of the idea of using CT's in aids for decision problems. Indeed, that approach has the following two limitations. First, it requires as input information *all* the indifference set I and *cannot* process any other input information, then the given information is processed by a *fixed*, specified clustering technique, due to

[Kruskal a,b].

Recently [G.G. Roy] has offered an approach which to a certain extant goes beyond the above-mentioned first limitation of [Rivett].

The method in [Mond and Rosinger c] presents an approach which eliminates both of the limitations in [Rivett] in the following way. First, it becomes possible to give as input information any subset $D \subset X \times X$ such that

$$D \subset I \cup P \cup Q \cup R \tag{75}$$

where I is the indifference set, while $P, Q, R \subset X \times X$ are respectively the strict preference set, large preference set and incomparability set, as defined in [B. Roy]. Then, it will also become possible to use a variety of CT's in order to obtain a cluster in X which corresponds to the most desirable feasible decisions.

The *first advantage* of this extension is in the significantly increased *freedom* the DM has in providing the input information $D \subset X \times X$. Indeed, under very general assumptions the family of sets $(I, P, P^{-1}, Q, Q^{-1}, R)$ is a *partition* of $X \times X$. Morevoer, the larger the number of objectives, the larger the size of R when compared to the other sets in the above partition. In addition it is well-known that, especially in such cases, the easiest thing for the DM is to specify pairs $(x,y) \in R$, even if not all of them. Therefore, the relaxation of the condition $D = I$ required in [Rivett] by the new condition (75) is particularly significant and useful.

The *second advantage* of the extension of the method in [Rivett] presented in [Mond and Rosinger c] is the following. As mentioned in Section 0, none of the *solution concepts* presently in use in decision problems with conflicting objectives, such as for instance Pareto optimum, utility function optimization, etc., are considered to be completely satisfactory. Indeed, one of the main complaints against them comes from their *a priori* nature, i.e., they define solution concepts prior to and thus independently of the particular *decision processes* at hand. In this way they introduce an *a priori bias* which cannot easily be balanced during the particular decision process to which they are applied. A well-known way of trying to avoid this inconvenience is offered by the on-line *interactive* methods in which the DM gets involved in an iterative, step-by-step procedure leading to a solution. In this way the *solution concept* is no more given a priori but rather follows in an *algorithmic, operative* way, part of which consists of the DM's *local* idea of solution or improvement at each step of the iteration. However, as experienced in practice, a major disadvantage of the on-line interactive methods is that they may lead to a large number of iterations thus straining and sometimes confusing the DM. In this connection the method presented in [Mond and Rosinger c] with the possiblity of choosing *various* CT's for processing the input information D has the *double advantage* of keeping the resulting solution concepts on an algorithmic, operative level without at the same time subjecting the DM to hard, on-line demands. Indeed, this time the DM can simply choose in an a priori way a certain CT, judging it according to its prior performance on *similar or test* problems.

After that, the chosen CT can be applied to the input information D without the DM's participation in the solution process.

REFERENCES

EVERITT, B. : Cluster Analysis (Second Ed.), Halsted, London, 1980.

FRANK, M., WOLFE, P. : An Algorithm for Quadratic Programming. Naval Research Logistic Quarterly, 3, 1956, 95 - 110.

GEOFFRION, A.M., DYER, J.S., FEINBERG, A. : An Interactive Approach for Multicriterion Optimization with an Application to the Operation of an Academic Department. Management Sci., 19, 1972, 357 - 368.

GERSHON, M., DUCKSTEIN, L. : An Algorithm for Choosing of a Multiobjective Technique. Paper presented at the Fifth International Conference of Multiple Criteria Decision Making, Mons, Belgium, August 1982.

GORDON, A.D. : Classification Methods for the Exploratory Analysis of Multivariate Data. Chapman and Hall, London, 1981.

GRUBER, J. : Introduction: Towards Observed Preferences in Econometric Decision Models. In J. Gruber (Ed): Econometric Decision Models. Springer Lecture Notes in Economics and Mathematical Systems, vol. 208 1983, pp 1 - 9.

HWANG, C.L., YOON, K. : Multiple Attribute Decision Making, Methods and Applications, A State of the Art Survey. Lecture Notes in Economics and Mathematical Systems, vol. 186, Springer, New York, 1981.

KENDALL, D.G. : The Recovery of Structure from Fragmentary Information, Phil. Trans. Royal Soc., A279, 1975, 547 - 582.

KRUSKALL, J.B. : Multidimensional Scaling by Optimizing a Goodness of Fit to a Nonmetric Hypothesis. Psychometrica, 29, 1964, 1 - 27.

KRUSKALL, J.B. : Nonmetric Multidimensional Scaling: A Numerical Method. Psychometrica, 29, 1964, 115 - 129.

MOND, B., ROSINGER, E.E. : Interacting Weights in Multiple Attribute Decision Making. Submitted for publication.

MOND, B., ROSINGER, E.E. : Multiple Attribute Decision Making: A Linear Programming Approach. Submitted for publication.

MOND, B., ROSINGER, E.E. : Multiple Criteria Decision Making : A Clustering Technique Approach. Submitted for publication.

RIVETT, P. : The Use of Local-Global Mapping Techniques in Analysing Multi Criteria Decision Making. In: Fanded, G., Gal, J. (Eds.), Multiple Criteria Decision Making: Theory and Applications. Lecture Notes in Economics and Mathematical Systems, vol. 177, Springer, New York, 1980.

ROSINGER, E.E. : Interactive Algorithm for Multiobjective Optimization (extended abstract) In: Fanded, G., Gal, J. (Eds.): Multiple Criteria Decision Making: Theory and Applications, Lecture Notes in Economics and Mathematical Systems, vol. 177, Springer, New York, 1980.

ROSINGER, E.E. : Interactive Algorithm for Multiobjective Optimization. JOTA, 35,3, 1981, 339 -365. Errata Corrige, JOTA, 38,1, 1982, 147 - 148.

ROY, B. : Partial Preference Analysis and Decision Aid: The Fuzzy Outranking Relation Concept. In: Bell, D.E., Keeney, R.L., Raifa, H. (Eds.): Conflicting Objectives in Decisions. J. Wiley, New York, 1977.

ROY, G.G. : The Use of Multidimensional Scaling in Policy Selection. J. Oper. Res. Soc., 33, 1982, 239 - 245.

SAATY, T.L. : A Scaling Method for Priorities in Hierarchical Structures. J. of Math. Psychology, 15,3, 1977, 234 - 281.

SAATY, T.L. : Ratio Measurement, Intangibles and Complexity in Systems. Paper presented at the Fifth International Conference on Multiple Criteria Decision Making, Mons, Belgium, August 1982.

SCHIFFMAN, S.S., REYNOLDS, M.L., YOUNG, F.W. : Introduction to Multidimensional Scaling: Theory, Methods and Applications. Acad. Press, New York, 1981.

SPÄTH, H. : Cluster Analysis Algorithsm for Data Reduction and Classification of Objects. J. Wiley, New York, 1980.

SPRONK, J. : Interactive Multiple Goal Programming, Applications to Financial Planning. Nijhoff Boston, 1981.

STREUFF, H., GRUBER, J. : The Interactive Multiobjective Optimization Method by E.E. Rosinger: A Computer Program and Aspects of Applications. In J. Gruber (Ed): Econometric Decision Models. Springer Lecture Notes in Economics and Mathematical Systems, vol. 208, 1983, pp. 334 - 364.

WALLENIUS, H. : Optimizing Macroeconomic Policy: A Review of Approaches and applications. European J. Oper. Res., 10, 1982, 221 - 228.

WALLENIUS, J. : Comparative Evaluation of Some Interactive Approaches to Multicriterion Optimization. Management Sci., 21, 1975, 1387 - 1397.

WOLFE, P. : The Simplex Method for Quadratic Programming. Econometrica, 27, 1959, 382 - 398.

WOLFE, P. : Convergence Theory in Nonlinear Programming. In: Abadie, J. (ed): Integer and Nonlinear Programming. North Holland, Amsterdam, 1970.

ZIONISTS, S., WALLENIUS, J.: An Interactive Programming Method for Solving the Multiple Criteria Problem. Management Sci., 22, 1976, 652 - 663.

THE ANALYTIC HIERARCHY PROCESS: EXPERIMENTS IN STABILITY

A.G. Lockett, B. Hetherington
Manchester Business School

ABSTRACT

The Analytic Hierarchy Process has been applied to many problems, and major claims made about its value. In this paper we look at how robust the procedure is in real life situations. First of all the method is outlined and the principles illustrated using a small numerical example, and a real life case study. This is followed by the results of a series of experiments using groups of people over time, to see how their subjectivity alters. Finally some simulations are carried out on data obtained in practice. The results indicate the individual data stability is high, and point to the more important factor of initial model specification.

INTRODUCTION

The ideas concerning Multi Criteria Decision Making have largely arisen in the last decade, and many of the approaches have been developments of mathematical programming methods. In the early stages this resulted in goal programming and a host of related techniques. Although having great mathematical complexity, conceptually they are very close to the original ideas embedded in linear programming i.e. optimising of objective functions in multi-dimensional spaces. The resulting procedures have meant complex mathematical computer packages of enormous power to cope with the daunting combinatorial problems they encounter.

However, none of these approaches has taken a step back from the original models to consider the problem from the users point of view and his or her world. The implicit assumption has been that because L.P. et al have been reasonably successful, a continuation along that road would be similarly rewarding. Even though there have been some examples of implementation, the general picture is not so bright. We have many more models than applications, and it is this which is our primary interest here - the application of mathematics to help decision makers.

In parallel with the programming type of modelling there has been a novel methodology, taken from a completely different standpoint. It is called the Analytic Hierarchy Process (AHP) and has been developed by

Saaty (1). Only a condensed version will be presented in this paper as we do not have the space to do justice to the method. However, it is hoped that we will interest some of you to delve into the references.

In the next section we will present an introduction to AHP including a small numerical example. The mathematics is kept to a minimum, because it is not the most important part of the methodology, in contrast to other procedures. Following that, some real life results will be presented and analysed, and their implications discussed. The fundamental problems of dealing with subjective data will be addressed - for all types of multi objective optimization. Some experiments and simulation results will then be presented which give insights into the data stability, and hence implications for modellers in this area.

AN OUTLINE OF AHP

Multiple criteria decision making implies that one has to weight differing factors. Once we are passed the simple maximising profit or minimising cost models this is inevitable. But how do we make such judgements as the cost of a car versus its size, or location of a house versus the environment? Most methods assume that the data is given - you just ask someone.

The Hierarchy

The AHP takes a different view and looks at how people make comparisons between objectives and things. One central premise of the methodology is that human beings cannot deal with complex decisions all at once, and break such problems down into a series of simpler and related decisions i.e. a hierarchial structure. As stated previously we cannot really do justice to the method here and a lot of the richness will be lost in these few pages. There is only enough space to give a flavour of the methodology. Let us take a simple example - say that of choosing a type of holiday. How do we set about it and what are the factors that we take into account? First of all we have to list the possible options or choices, as exemplified in Figure 1.

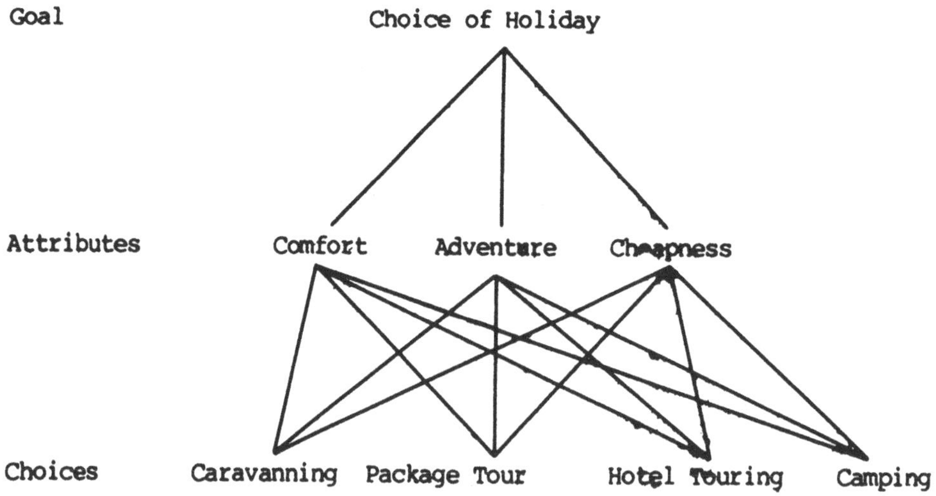

FIGURE 1: HIERARCHY FOR HOLIDAY CHOICE PROBLEM

Before we can make our decision(s) we also have to list the factors or attributes that affect our choices, and again these are shown in Figure 1. In this simple example the choices are camping, caravaning, package tour, and hotel touring. The attributes are taken to be comfort, adventure and cheapness, and are what the author thinks are important. Other decision makers probably have completely different sets of attributes. If we look at them they are seen to be very different. Comfort is an attempt to measure the varying degrees of comfortableness of the holiday, e.g. it is probably more enjoyable to be on a hotel bed (air conditioned) than to sleep in a tent. Adventure takes into account that some of the choices may provide a more exciting type of holiday. The third attribute looks at the most common attribute - money. Here we have turned it on its head and worked in terms of cheapness rather than cost. To the purist we are already in trouble. To the realist we are showing a way forward. All the attributes we are dealing with are equally "fuzzy", and hence our small concern with the above problem. Most multi objective problems have an inherent fuzziness, i.e. a blurring, about the data that is being used. More importantly there is also an ambiguity about the goals of the exercise (Lockett and Hetherington (2)). Therefore we are recognising the real environment within which modellers have to work.

Overall we wish to make a choice between a group of options, and those options have a series of attributes associated with them. The problem structure is shown in Figure 1, and is an illustration of a simple hierarchy. Our methodology splits off the attributes from the choices. In the next phase we compare the attributes with each other, i.e. we try to weigh them against each other. We have only illustrated a simple form of 2 stage hierarchy, but the principle can be applied to many levels (e.g. see Wind and Saaty (3)). Again the basic idea is to break down the problem into various levels of decision, into a series of simpler related stages. Unfortunately mathematics does not help us here, and model structuring is an art, not a science. However there are many examples of more complex problems and associated solutions in the literature. We now return to the problem of choosing weights.

Pairwise Comparison

We wish to compare a number of things, and the AHP methodology does this by a series of pairwise comparisons i.e. only two are taken at a time. There is a large amount of psychological literature which underpins this reasoning - and hence makes the basis of this approach different (not necessarily better) than the more orthodox ones. It is attempting to fit models to people, rather than people to models. Again time and space limitations make us assume that this is well known. For the more critical reader, time spent in the literature may prove mind stretching as well as rewarding.

Suppose we wish to compare n objects which are A1, An, and wish to find their weights (assumed to belong to a ratio scale) which are w1, w2, wn. Our method asks us to give a series of ratios of paired comparisons i.e. A1 / A2 = a12, etc, giving a pairwise comparison matrix:

A =

	A1	A2		An
A1	a11	a12	. . . .	a1n
A2	a21	a22		.
	.			.
	.			.
	.			.
An	an1	. . . .		ann

The aij's are subjective estimates of how we feel about the two items in question i.e. data. In most instances the problem is simplified by assuming that the diagonals are 1, and that aij = i/aji. But we want to know what are the underlying wi's - the weights. These are part of the data because aij = wi/wj. If the person giving the original subjective data was totally consistent we would not have a problem. For example, consider the simple example comparison matrix below:

	C1	C2	C3
C1	1	2/1	3/1
C2	1/2	1	2/3
C3	1/3	3/2	1

The data is entirely consistent. If we take w1 to be 1, then w2 = 1/2, w3 = 1/3, and we could normalise the results to produce the answer we require i.e. w_1 = Ø.545, w_2 = Ø.273, w_3 = Ø.182. But in real life people may not be totally consistent - some of their preferences may be vague. That is usually found, and it is unlikely that we will be exact. Human nature and the problems under consideration make this the realistic case, and a more likely comparison matrix would be

	C1	C2	C3
C1	1	2/1	3/1
C2		1	1/2
C3			1

Note that we have assumed that only the entries above the diagonal are necessary. Now we have some inconsistency. If the first two entries are correct i.e. 2/1, and 3/1, then C2/C3 should be 2/3 as previously. But why should that be incorrect, and not one of the other entries. One could ask the respondent to redo one or more of the entries - and produce consistency. However, then we are back to simple weightings and the matrix is not needed. The very essence of this method is that it allows such fuzziness to occur and work with it rather than driving it out. That being so, how do we find the 'best' weightings? This is where the mathematics of optimization comes in. We wish to find the weightings which are most reasonably interpreted from the series of pairwise comparisons.

For the consistent case we can show that

$$Aw = nw$$

and hence given A, to find w (the eigenvector), we solve $(A - nI)w = 0$. It can be shown that all the eigenvalues of A are zero except $\lambda max = n$.

In the 'inconsistent' case, the ratio scale is not known, and we have estimates of the ratios, which are not necessarily consistent. The problem Aw=nw becomes $Aw' = \lambda max\ w'$ where λmax is the largest eigenvalue of A (see Saaty (4)). Again we find w' and λmax (>n). For our inconsistent example this gives us the result $w_1 = .547$, $w_2 = .189$, $w_3 = .263$, $\lambda max = 3.136$.

Hence we are able to easily find the underlying weightings associated with the paired comparisons - without forcing the decision maker to be exactly consistent. We are also using a lot of redundant data which is required by this method i.e. n(n-1)/2 pieces of data compared to n. This means that inaccuracies are much less important in our method, but this is gained at the expense of extra work/input.

We can also show that the more inconsistency the larger λmax and this can be used as a basis for estimating the degree of consistency. Saaty uses an index $(\lambda max - n)/n$, but we prefer to use:- $(\lambda r - \lambda max) \times 100/(\lambda r - n)$ where λr = the average eigenvalue obtained by random input using the same scale. Therefore the λr's vary with the size of the matrix, and have to be produced using simulation techniques. It would be pleasant to report that they could be derived analytically, but unfortunately at this point in time it is not possible. The above index has the property that for an entirely consistent input it is 100%, and for random it is 0%. Linear interpolation is assumed in between. At this stage in our understanding that is all that can be said - it is a measure, and a rough indicator. We cannot say that a more consistent decision maker is better than one who is less consistent.

The main component we have not so far touched on is the scale to be used. There are many arguments on this subject, and the psychology literature is where it is usually found. Most of the research suggests that the best scale should have the magic number 7 ± 2 points. For the examples seen later in this paper we have used the psychological scale shown in Figure 2.

It should be remembered that this is a ratio scale, and the numbers express the preferences of the decision makers. Each person may make his own interpretation of the scale - and in fact defines it by the judgements. This is completely different to the normal multi objective programming which are usually based on additive utilities. There is a debate in the literature about this approach which goes well beyond mathematics, and involves the various philosophies of different disciplines. All we will say at this stage is that at present the

INTENSITY OF IMPORTANCE	DEFINITION	EXPLANATION
Ø	NOT COMPARABLE	IT IS NOT MEANINGFUL TO COMPARE THE TWO ACTIVITIES
1	EQUAL IMPORTANCE	TWO ACTIVITIES CONTRIBUTE EQUALLY TO THE OBJECTIVE
3	WEAK IMPORTANCE OF ONE OVER ANOTHER	THERE IS EVIDENCE FAVOURING ONE ACTIVITY OVER ANOTHER, BUT IT IS NOT CONCLUSIVE
5	ESSENTIAL OR STRONG IMPORTANCE	GOOD EVIDENCE AND LOGICAL CRITERIA EXIST TO SHOW THAT ONE IS MORE IMPORTANT
7	DEMONSTRATED IMPORTANCE	CONCLUSIVE EVIDENCE AS TO THE IMPORTANCE OF ONE ACTIVITY OVER ANOTHER
9	ABSOLUTE IMPORTANCE	THE EVIDENCE IN FAVOUR OF ONE ACTIVITY OVER ANOTHER IS OF THE HIGHEST POSSIBLE ORDER OF AFFIRMATION
2,4,6,8	INTERMEDIATE VALUES BETWEEN THE TWO ADJACENT JUDGEMENTS	
RECIPROCALS OF ABOVE NON-ZERO NUMBERS	IF ACTIVITY I HAS ONE OF THE ABOVE NON-ZERO NUMBERS ASSIGNED TO IT WHEN COMPARED WITH ACTIVITY J, THEN J HAS THE RECIPROCAL VALUE WHEN COMPARED WITH I	

FIGURE 2: RATIO SCALE USED IN THE EXAMPLES

methodology is well liked by decision makers and is apparently gaining more ground than some other methods. We are now in a position to develop the overall procedure.

Methodology

The first step is to develop the structure of the hierarchy. Once this is done the factors at the highest level of the hierarchy are compared, using the pairwise procedure described above. Let us assume that this produces the following comparison matrix as shown in Table 1.

	Cheapness	Comfort	Adventure
Cheapness	1	6/1	2/1
Comfort	1/6	1	1
Adventure	1/2	1	1

Eigenvalue = 3.136 Consistency 88.3%

TABLE 1: ATTRIBUTE COMPARISON MATRIX

From which we can derive an eigenvector wa = (wa_1, wa_2, wa_3) = (0.63, 0.15, 0.22) and eigenvalue max = 3.136. First of all let us consider the eigenvalue. If there had been 100% consistency max = 3, and the difference between the two gives us a consistency of 88.3% using our formula. Let us assume that we are satisfied with this result for the present.

The next stage is to compare for each of the attributes the series of choices and we may end up with the following matrices as shown in Tables 2 (a,b,c): Choice Comparison Matrices

Attribute:- Cheapness

	Camping	Caravan	Hotel Touring	Package Tour
Camping	1	7/1	9/1	7/1
Caravan	1/7	1	5/1	4/1
Hotel Touring	1/9	1/5	1	5/1
Package Tour	1/7	1/4	1/5	1

Eigenvalue 4.701 Consistency 74.0%

TABLE 2(a): CHOICE COMPARISON MATRIX

Attribute:- Comfort

	Camping	Caravan	Hotel Touring	Package Tour
Camping	1	1/5	1/9	1/6
Caravan	5/1	1	1/8	1/3
Hotel Touring	9/1	8/1	1	7/1
Package Tour	6/1	3/1	1/7	1

Eigenvalue 4.413 Consistency 84.7%

TABLE 2(b): CHOICE COMPARISON MATRIX

Attribute:- Adventure

	Camping	Caravan	Hotel Touring	Package Tour
Camping	1	6/1	7/1	8/1
Caravan	1/6	1	4/1	6/1
Hotel Touring	1/7	1/4	1	3/1
Package Tour	1/8	1/6	1/3	1

Eigenvalue 4.335 Consistency 87.6%

TABLE 2(c): CHOICE COMPARISON MATRIX

It can be seen that for each of the attributes the choices vary in their relative attractiveness. In each instance the consistency is above 70% and we have assumed that it is satisfactory.

The eigenvector of the attribute matrix (Table 1) is given by

	Choice of Holiday	
	Cheapness	.63
wa =	Comfort	.15
	Adventure	.22

The matrix of eigenvectors of the choice comparisons matrices is given by

		Cheapness	Comfort	Adventure
	Camping	.67	.04	.66
	Caravan	.20	.10	.21
wb =	Hotel Touring	.09	.69	.08
	Package Tour	.04	.17	.05

The final stage works out the overall preference by weighting these eigenvectors with the results from the earlier comparisons i.e. wa^t = (.63, .15, .22) a linear weighted sum utility function which gives an overall weighting of 0.58, 0.18, 0.7 and 0.07 respectively. Hence in this instance camping is the most preferred option.

We have illustrated above the simplest form of the method in operation. Much more complex hierarchies can be dealt with, and Saaty et al (5) (6) (7) have developed a mathematical approach to the general system. However, we feel that this can obscure the inherent assumptions in the methodology which require clear understanding if it is to be used most effectively.

What we have done is akin to first cutting up a cake into attribute slices. Each of these is then divided up amongst the choices, and finally these separate little pieces are added together as illustrated in Figure 3.

The only (but important) part that mathematics takes in the procedure is that of calculation of the eigenvector and eigenvalue. But what is important in human (or behavioural) terms about the method is that it copes with fuzziness and ambiguity in a way that is acceptable to people. It is very "user friendly". Also it has one other advantage over many other methods in that it does not take its weightings directly but in a sense "averages" the input data. Hence single "errors" are not totally disturbing.

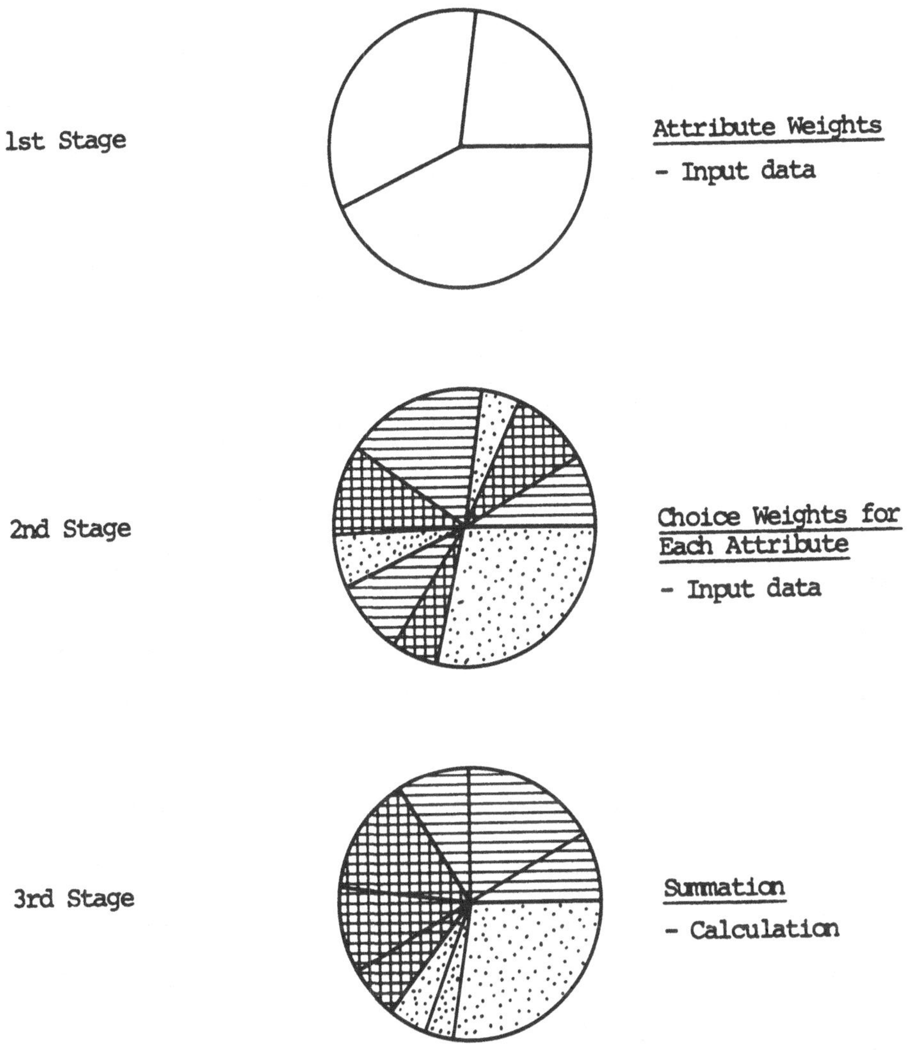

FIGURE 3: DIAGRAMMATIC REPRESENTATION OF CHOICE ATTRIBUTE NUMERICAL STAGES

SOME RESULTS

We have used AHP on a number of problems and the results are fully reported elsewhere ((8) (9)). In the cases under discussion groups of people have been the focus of our study, because they have been the decision making body rather than any one individual. It could be argued that multi objective optimisation is mainly the province of groups rather than single decision makers, and hence our interest is not without support.

Group Data

Our results indicate that the subjective data from the people on varying types of problem have some stability, in that the range variation is reasonably predictable. For example in one case of investment appraisal, a group of four people defined the following basic structure of choices and attributes as shown in Table 4. Their intermediate and final overall results are presented in Table 5. A glance at the data suggests that the variation between the different peoples subjective answers is related to the average solution. Using this idea, a linear regression was performed of the range (i.e. difference between largest and smallest result) and the arithmetic mean of the final choice weights. The results of this analysis are presented in Table 6 and shown as case C, along with similar analysis for the other unrelated applications of the methodology. The unadjusted R^2

values are extremely high and all are very similar in value. The slopes of the derived lines are also close and there are no significant differences between the cases. If we plot the data of the combined series of R/A, the results compare to the normal distribution as shown in Figure 4.

CHOICE		ATTRIBUTES	
No.	Name	No.	Description
1	Sanitary Towels	1	Return on capital employed
2	Soft drink manufacture	2	Best fit on size
3	Medical cotton wool	3	Stability
4	Security devices	4	Competitiveness
5	Property	5	Growth
6	Electrical/Electronic distribution	6	Market share
7	Regional breweries	7	Quality of management
8	Textile imports	8	Technology

TABLE 4: FINAL CHOICES AND ATTRIBUTES FOR INVESTMENT APPRAISAL CASE

```
  ABS. FREQ.

    10                           *
      -                          *
      -                          *
      -                          *
      -                         **
     5                          **
      -                         ** *
      -                         ****
      -                        ******
      -                       ********
       ------:---------:---------:---------:---------:------
         -3.00E+00 -1.00E+00  1.00E+00  3.00E+00  5.00E+00

                             R/A
MEAN       =  1.0499E+00
STD. DEV. =  3.1978E-01
SAMPLE SIZE =    29
```

FIGURE 4: DISTRIBUTION OF (RANGE/AVERAGE WEIGHT)
FOR COMBINED DATA

	PERSON A		PERSON B		PERSON C		PERSON D		AVERAGE	
ATTRIBUTE	ACT.	RNK	ACT.	RNK	ACT.	RNK	ACT.	RNK	ACT.	RNK
RET. ON CAPITAL	8.2	5	32.1	1	19.9	1	31.7	1	23.Ø	1
BEST FIT ON SIZE	21.8	2	4.9	7	2.7	8	1.9	8	7.7	7
STABILITY	7.2	6	13.8	3	13.7	4	11.5	4	11.6	5
COMPETITIVENESS	12.7	4	9.4	5	6.5	2	23.9	2	13.1	3
GROWTH	25.9	1	16.3	2	24.5	3	15.6	3	2Ø.6	2
MARKET SHARE	18.9	3	13.6	4	4.1	6	3.2	6	9.9	6
MANAGEMENT QUAL.	3.8	7	8.4	6	24.9	5	1Ø.7	5	11.9	4
TECHNOLOGY	1.5	8	1.5	8	3.7	7	2.Ø	7	2.2	8
INVESTMENT AREA										
ELECTRONIC DIST	21.3	1	9.5	5	8.9	6	11.7	5	12.6	5
MED. COTTON WOOL	13.7	2	8.3	6	23.8	1	2Ø.6	2	16.6	2
REG. BREWERIES	13.6	3	17.Ø	3	11.1	4	13.3	4	13.8	4
SOFT DRINK MFG	12.2	4	15.8	4	13.7	3	4.Ø	6	11.4	6
SECURITY DEVICES	11.5	5	22.9	1	8.5	7	15.Ø	3	14.5	3
TEXTILE IMPORTS	1Ø.9	6	4.Ø	7	3.6	8	3.2	7	5.5	7
SANITARY TOWELS	1Ø.7	7	19.2	2	21.4	2	29.3	1	2Ø.2	1
PROPERTY INVESTMT	6.1	8	3.3	8	9.Ø	5	2.9	8	5.4	8

TABLE 5: ATTRIBUTE AND FINAL CHOICE RESULTS

Case	No. of Values	Equation	Unadjusted R^2
A	11	R=1.0602 + 0.8530A	0.6874
B	10	R=3.3918 + 0.7622A	0.7262
C	8	R=2.4656 + 0.7322A	0.6557
A+B+C	29	R=2.2998 + 0.7778A	0.6969

TABLE 6: REGRESSION EQUATIONS FOR THE SEPARATE AND COMBINED CASES

(R = range, A = average weight)

What can we infer from our findings? From a set of completely different experiments, people and environments, the derived variations are remarkably alike. We expected the range to increase with the average weight, but not in the simple manner shown here. If our results were found to be obtained in general, it would be of considerable use in MCDM modelling. The size of uncertainties could bring into question much of the mathematical modelling that is currently taking place.

At this stage of our understanding an exact explanation for our findings is not possible, and we can only speculate. One possible suggestion is that the problems are all of a similar type - where groups of people have already found it difficult to come to an answer, and the variation is "typically" what you would expect. In other words, on real problems this is the variation that will not go away. Modelling like ours may bring this out and thereby helpful in the process, but models will not make it any easier. It may be that many approaches obscure the differences by going for cosmetic agreements - that is why they may be useful in the process of getting decisions accepted, not for their modelling powers. At this time we can only hazard a guess, but our preliminary findings suggest caution.

However, we accept the limitations in our experimental design and have looked at some possible problem areas. Two of these are the variation of subjective data over time and individual stability and these are investigated in the next section.

Repeatability

One problem that affects all MCDM is the problem of the possbile changing nature of subjective data i.e. will a person give the same numbers tomorrow as today? How stable are one persons solutions? To look into this more closely a series of experiments have again been done using groups of people. The results are fully reported in Lockett et al (10) and only part of them will be presented here.

Six Ph.D students in a Business School were being taught the A.H.P. and wanted a problem to use that would interest the whole group. It was decided to use the "choice of a daily newspaper". Six newspapers were considered for the set, three of which were the "tabloid" type, and the other three were from the "quality" range. Each individual was asked to express his/her choice using a two level hierarchy of attributes and choices and this was done on three consecutive days. The group was presented with differing sets of newspapers on the three occasions, and each choice set was considered as a separate entity. In this experiment the group used the same common set of attributes which they had derived. The attribute results are shown in Table 7.

On examination of the attribute weights a high degree of stability is evident over the three days and although there are minor variations this could probably be due to changes in newspaper quality each day. For the problem in question there is a stable attribute set for each person, irrespective of the choice set.

On discussing the problem for a few weeks, 3 of the individuals concluded that the original attributes were not sufficient. They requested another try. The result was a rerun of the third exercise using the whole choice set (6 newspapers) with a new set of attributes as shown in Table 8.

SIX INDIVIDUAL'S ATTRIBUTE WEIGHTS FOR THREE SETS OF CHOICES

	INDIVIDUAL 1			INDIVIDUAL 2		
ATTRIBUTE	DAY 1	DAY 2	DAY 3	DAY 1	DAY 2	DAY 3
CHEAPNESS	4.52	3.86	4.80	4.30	3.47	3.94
PRESENTATION	11.70	16.46	11.28	5.98	5.48	11.47
POLITICAL COVERAGE	46.87	48.36	46.49	53.61	51.19	37.01
SPORTS COVERAGE	2.58	3.34	3.04	1.68	1.80	11.55
GENERAL NEWS	21.67	18.45	20.01	17.22	19.03	13.07
FEATURES	12.67	9.53	14.38	17.22	19.03	21.93

	INDIVIDUAL 3			INDIVIDUAL 4		
ATTRIBUTE	DAY 1	DAY 2	DAY 3	DAY 1	DAY 2	DAY 3
CHEAPNESS	19.08	25.10	24.45	2.19	2.20	2.57
PRESENTATION	17.60	9.71	8.54	47.72	46.08	43.33
POLITICAL COVERAGE	29.19	36.49	38.25	8.11	8.71	11.27
SPORTS COVERAGE	7.63	5.94	6.12	15.09	20.54	21.39
GENERAL NEWS	14.06	11.21	11.71	13.44	11.23	10.72
FEATURES	12.94	11.56	11.47	13.44	11.23	10.72

	INDIVIDUAL 5			INDIVIDUAL 6		
ATTRIBUTE	DAY 1	DAY 2	DAY 3	DAY 1	DAY 2	DAY 3
CHEAPNESS	2.67	2.85	3.07	4.14	5.92	5.25
PRESENTATION	20.30	24.80	28.61	10.35	16.00	14.03
POLITICAL COVERAGE	36.60	32.67	28.61	28.08	40.17	38.62
SPORTS COVERAGE	3.69	3.39	3.07	21.08	16.00	14.03
GENERAL NEWS	9.68	10.00	9.51	26.00	16.00	14.03
FEATURES	27.06	26.30	27.13	10.35	5.92	14.03

TABLE 7: NEWSPAPER CHOICE

The new weights are presented alongside the original details. Inspection of the table reveals that most of the original attributes were also incorporated into the new set, although their exact definition changed. For example political coverage was still included, but it did not encompass the new attribute of ideological fit. It was now taken to mean how well politics was covered by the newspaper, rather than how the person reacted to it. In all cases except one, the original attributes in the larger set received a smaller weighting. The final choice results for the 3 people are presented in Table 9.

Again there is a high degree of stability, and using a very different set of attributes the method gives the same choice results. It should also be remembered that the two experiments were conducted with a two month interval in between.

It appears that from a very different attribute set, the final choices remain fairly constant. This could be viewed in two totally opposite ways. One is to say that, if this were the case in similar types of decision situtation, the model is producing the "right answer". The opposite view is that each participant had a known answer and the method allowed him or her to derive the result they wanted.

THREE INDIVIDUAL'S ATTRIBUTE WEIGHTS FOR ONE SET OF CHOICES

	INDIVIDUAL 1		INDIVIDUAL 2		INDIVIDUAL 3	
ATTRIBUTES	9 ATT.	6. ATT	9 ATT.	6 ATT.	9 ATT.	6 ATT.
CHEAPNESS		4.80		2.57		3.07
IDEOLOGICAL FIT	22.66		11.12		33.70	
SPORTS NEWS	1.26	3.04	26.02	21.39	1.30	3.07
POLITICAL COVERAGE	18.46	46.49	6.38	11.27	14.87	28.61
BUSINESS NEWS	15.54		14.80		10.50	
GENERAL NEWS & INFO	10.49	20.01	5.77	10.72	3.18	9.51
FEATURES	6.76	14.38	5.20	10.72	15.76	27.13
ARTS	5.74		5.86		3.15	
PRESENTATION	6.34	11.28	5.66	43.33	7.33	28.61
INTELLECTUAL LEVEL	12.75		19.19		10.21	

TABLE 8: NEWSPAPER CHOICE

However the results are viewed it would seem that time does not present a problem. Unfortunately it has not been possible thus far to re-run any of the "live" projects since the people taking part have, after one "run", returned to their working environment satisfied with their results obtained in their particular project.

	INDIVIDUAL 1		INDIVIDUAL 2		INDIVIDUAL 3	
CHOICES	6 ATT.	9 ATT.	6 ATT.	9 ATT.	6 ATT.	9 ATT.
TIMES	42.00	40.33	39.30	37.09	28.88	32.28
GUARDIAN	26.20	27.33	27.23	34.39	34.58	42.52
TELEGRAPH	15.93	18.09	19.86	17.51	17.81	10.40
MIRROR	7.23	7.16	4.56	4.97	8.47	6.17
MAIL	3.75	4.30	5.85	3.57	5.83	5.30
SUN	4.89	2.80	3.20	2.46	4.43	3.33

TABLE 9: NEWSPAPER CHOICE

Individual Stability

The last section dealt with the subjective data that has been obtained with groups of people on real problems, and on a series of experiments over time. In general they show the potential of the method as a decision making aid.

For the single decision maker it is of importance to know how stable are the results, and what type of errors cause problems. It would be useful if mathematical procedures could be developed to cover this problem. Some work has been done e.g. Vangas (11), but at this point in time the mathematics in general is intractible. We have resorted to simulation, using our earlier real life data.

Some Simulations

The results we present here are based on the investment case described earlier in this chapter. Similar solutions have been obtained using the other experiments, but they are too voluminous to present here. However, our findings apply to all the cases we have so far investigated.

To get a general feel for the robustness of the model, three small data change analyses have been performed on the attribute input data. This will have a larger impact than changes in the later data. In the first instance for each person the largest input number was identified (regardless of sign) and gradually reduced to 1, as long as the consistency kept above zero. The results for the four people are presented in Table 10, which shows that for a single data change there is little or no effect. Hence one small error of this nature is unimportant in practice.

NO.	ATTRIBUTE	PERSON A ORIG.	PERSON A CHNG.	PERSON B ORIG.	PERSON B CHNG.	PERSON C ORIG.	PERSON C CHNG.	PERSON D ORIG.	PERSON D CHNG.
	DATA POINT CHNG.	5/8		1/8		1/2		7/8	
1	RET. ON CAPITAL	8.2	8.9	32.1	27.8	19.9	17.8	31.7	31.7
2	BEST FIT ON SIZE	21.8	21.2	4.9	5.7	2.7	4.2	1.9	1.4
3	STABILITY	7.2	7.9	13.8	13.9	13.7	14.4	11.5	11.8
4	COMPETITIVENESS	12.7	12.5	9.4	9.9	6.5	6.9	23.9	23.8
5	GROWTH	25.9	22.5	16.3	16.1	24.5	24.6	15.6	15.6
6	MARKET SHARE	18.9	19.1	13.6	13.7	4.1	4.0	3.2	3.4
7	MANAGEMENT QUAL.	3.8	4.5	8.4	9.3	24.9	24.5	10.7	9.5
8	TECHNOLOGY	1.5	3.4	1.5	3.6	3.7	3.6	2.0	2.8
	CONSISTENCY	84.7	72.6	81.6	67.2	74.3	71.3	60.2	59.4
	CHOICES								
	ELECTRONIC DSTBN	21.3	20.9	9.5	9.8	8.9	8.8	11.7	11.6
	MED. COTTON WOOL	13.7	14.1	8.3	8.5	23.8	23.7	20.6	20.6
	REG. BREWERIES	13.6	13.7	17.0	16.8	11.1	10.9	13.3	13.3
	SOFT DRINK MFG.	12.2	12.0	15.8	15.5	13.7	13.3	4.0	4.0
	SECURITY DEVICES	11.5	11.5	22.9	22.6	8.5	8.7	15.0	15.1
	TEXTILE IMPORTS	10.9	10.9	4.0	4.0	3.6	3.9	3.2	3.2
	SANITARY TOWELS	10.7	10.8	19.2	19.4	21.4	21.6	29.3	29.3
	PROPERTY INVSMT	6.1	6.1	3.3	3.4	9.0	9.1	2.9	2.9

TABLE 10: RESULTS OF CHANGING HIGHEST DATA POINT TO 1

Taking this one stage further the analysis has been repeated allowing the two largest numbers to be changed. Again the results are very stable as shown in Table 11 below. Hence we could have 2 large data errors and they can be accommodated without too much worry.

We could carry on with this line of argument and do much more statistical analysis, but for little gain. The simple simulations already presented give good indications of likely stability.

For the third simulation we took the single data case and assumed the worst result - a change of sign i.e. the input had inverted the comparison. In practice this happens from time to time - usually producing a low consistency and the error is then picked out. For this simulation we are assuming that the error is not noticed. There is only enough space to present the worst possible outcomes here i.e. when the largest numerical value has the wrong sign, and, of course, the typical result of a change would be less dramatic. The outcome of the simulation is presented in Table 12. In this case the effect on the weightings in the attributes is more marked, and the inconsistencies generally drop substanially. Nevertheless, the effect on the final choice weightings is not dramatic. The variations produced are much less than the variations between the experiments. Stability is still very high. Comparable results have been found for all the data analysis we have performed. In practice, the nature of the model, with its

		PERSON A		PERSON B		PERSON C		PERSON D	
	DATA POINT CHNG.	3/8 TO 6/7		3/8 TO 4/8		3/5 TO 4/8		4/7 TO 4/8	
NO.	ATTRIBUTE	ORIG.	CHNG.	ORIG.	CHNG.	ORIG.	CHNG.	ORIG.	CHNG.
1	RET. ON CAPITAL	8.2	9.5	32.1	31.5	19.9	22.3	31.7	30.7
2	BEST FIT ON SIZE	21.8	21.1	4.9	5.7	2.7	2.9	1.9	1.4
3	STABILITY	7.2	6.4	13.8	12.0	13.7	16.8	11.5	12.5
4	COMPETITIVENESS	12.7	13.2	9.4	8.3	6.5	4.7	23.9	18.3
5	GROWTH	25.9	26.6	16.3	16.3	24.5	17.6	15.6	16.6
6	MARKET SHARE	18.9	15.4	13.6	13.7	4.1	4.8	3.2	3.7
7	MANAGEMENT QUAL.	3.8	5.6	8.4	9.4	24.9	26.3	10.7	13.5
8	TECHNOLOGY	1.5	2.1	1.5	3.1	3.7	4.6	2.0	3.3
	CONSISTENCY	84.7	81.9	81.6	74.2	74.3	80.8	60.2	58.3
	CHOICES								
	ELECTRONIC DSTBN	21.3	22.2	9.5	9.9	8.9	8.7	11.7	11.8
	MED. COTTON WOOL	13.7	13.0	8.3	8.0	23.8	23.8	20.6	20.2
	REG. BREWERIES	13.6	12.6	17.0	17.2	11.1	10.8	13.3	13.1
	SOFT DRINK MFG.	12.2	12.3	15.8	15.8	13.7	13.1	4.0	4.9
	SECURITY DEVICES	11.5	11.8	22.9	23.1	8.5	8.8	15.0	15.6
	TEXTILE IMPORTS	10.9	11.2	4.0	4.0	3.6	3.7	3.2	3.3
	SANITARY TOWELS	10.7	10.8	19.2	18.6	21.4	21.6	29.3	29.0
	PROPERTY INVSTMT	6.1	6.1	3.3	3.4	9.0	9.5	2.9	3.0

TABLE 11: RESULTS OF CHANGING 2 HIGHEST DATA POINTS TO 1

		PERSON A		PERSON B		PERSON C		PERSON D	
	DATA POINT CHNG.	5/8 = -9		1/8 = -9		1/2 = -9		7/8 = -9	
NO.	ATTRIBUTE	ORIG.	CHNG.	ORIG.	CHNG.	ORIG.	CHNG.	ORIG.	CHNG.
1	RET. ON CAPITAL	8.2	10.0	32.1	19.0	19.9	14.2	31.7	29.1
2	BEST FIT ON SIZE	21.8	18.4	4.9	7.3	2.7	12.0	1.9	1.3
3	STABILITY	7.2	9.4	13.8	13.4	13.7	15.5	11.5	11.9
4	COMPETITIVENESS	12.7	11.8	9.4	10.5	6.5	8.2	23.9	22.7
5	GROWTH	25.9	15.7	16.3	19.7	24.5	21.8	15.6	14.7
6	MARKET SHARE	18.9	18.1	13.6	13.0	4.1	3.7	3.2	4.6
7	MANAGEMENT QUAL.	3.8	6.0	8.4	10.4	24.9	21.5	10.7	8.3
8	TECHNOLOGY	1.5	10.6	1.5	11.7	3.7	3.1	2.0	7.4
	CONSISTENCY	84.7	25.8	81.6	15.1	74.3	44.7	60.2	41.3
	CHOICES								
	ELECTRONIC DSTBN	21.3	20.3	9.5	11.0	8.9	9.9	11.7	11.6
	MED. COTTON WOOL	13.7	15.0	8.3	9.0	23.8	23.0	20.6	20.4
	REG. BREWERIES	13.6	13.6	17.0	16.1	11.1	10.6	13.3	13.0
	SOFT DRINK MFG.	12.2	11.3	15.8	14.7	13.7	12.0	4.0	4.1
	SECURITY DEVICES	11.5	12.1	22.9	22.6	8.5	9.5	15.0	16.1
	TEXTILE IMPORTS	10.9	10.3	4.0	4.0	3.6	4.0	3.2	3.1
	SANITARY TOWELS	10.7	11.4	19.2	19.2	21.4	21.7	29.3	28.9
	PROPERTY INVSTMT	6.1	6.1	3.3	3.4	9.0	8.3	2.9	2.8

TABLE 12: RESULTS OF CHANGING SIGN OF HIGHEST DATA POINT

inherent data redundancy, makes the effect of data errors a secondary problem. What is far more important is the model specification in the first instance, and the associated decision process that is employed.

CONCLUSION

In this chapter we have outlined a method of MCDM which is helpful in structuring the decision making problem. The cases and experiments presented, indicate its breadth of application and usefulness in many problem situations. Its robust methodology and user friendly characteristics make it a valuable aid for management.

However our group results suggest that optimisation may not be appropriate for many MCDM problems. We have shown that typically there is a large amount of inter-personal variation of the subjective data, and modelling will not necessarily resolve this. In our view it is much better to put more effort into understanding the process than into developing yet more mathematical complex models.

REFERENCES

1. Saaty, T.L., Analytical Hierarchy Process, McGraw Hill, New York, 1981.

2. Lockett, A.G., Hetherington, B., Subjective data and multiple criteria decision making, in Essays and Surveys on Multiple Criteria Decision Making, Ed., Hansen, Proceedings of the Vth International Conference on MCDM, Springer Verlag, New York, 1983.

3. Wind, Y., Saaty, T.L., Marketing applications of the analytic hierarcy process, Management Science, 26, 641, 1980.

4. Saaty, T.L., A scaling method for priorities in hierarchical structures, Journal of Math. Psychology, 15, 234, 1977.

5. Saaty, T.L., Wong, M.M., Projecting average family size in rural india by the A.H.P., J. of Math. Sociology, 9, 181, 1983.

6. Saaty, T.L., Conflict resolution and the falkland island invasions, Interfaces, 13, 68, Dec. 1983.

7. Saaty, T.L., Gholamnezhad, A.H. Oil prices: 1985 and 1990, Energy Systems and Policy, 5, No. 4, 303, 1981.

8. Lockett, A.G., Muhlemann, A.P., Gear, A.E., Group decision making and multiple criteria - a documented application, in Organisations: Multiple Agents with Multiple Criteria, Ed., Morse, Proceedings of IVth International Conference on MCDM, Springer-Verlag, New York, 1981.

9. Lockett, A.G., Gear, A.E., Muhlemann, A.P., A unified approach to the acquisition of subjective data in R & D, IEEE Trans. on Eng. Mgt., 29, No. 1, 11, Feb. 1982.

10. Lockett, A.G., Hetherington, B., Yallup, P., Subjective estimation and its use in MCDM, VIth Proceeding of VIth International Conference on MCDM, Cleveland, June 1984.

11. Vangas, L.G., Saaty, T.L., Inconsistency and rank preservation in MCDM, VIth Proceeding of VIth International Conference on MCDM, Cleveland, June 1984.

AN ALGORITHM TO SOLVE A TWO-LEVEL RESOURCE CONTROL PRE-EMPTIVE HIERARCHICAL PROGRAMMING PROBLEM

Subhash C. Narula
School of Business
Virginia Commonwealth University

Adiele D. Nwosu
Department of Mathematics
University of Nigeria

ABSTRACT

Recently the two-level resource control pre-emptive hierarchical programming problem has attracted much attention in the literature. However it is usually discussed under the misnomer 2-level linear programming problem. By taking into consideration the sequential nature of the decision making process in this problem, we develop an efficient algorithm to solve the problem. The proposed algorithm uses regular simplex pivots with a few modifications to select the variables to enter and leave the basis. The algorithm is illustrated with an example.

INTRODUCTION

Consider an organization in which control over the decision variables is partitioned among a hierarchy of independent decision makers; and where decisions are not made simultaneously and in concert but in sequence down the hierarchy (e.g., the hierarchy of federal, state and local governments in a democracy). As the objective functions and constraints of the system are, in general, functions of all decision variables, the independent actions of lower level decision makers often impact adversely on the objective function values of higher level decision makers. However, the sequential nature of the decision process enables a higher level decision maker to use his superior rank to unilaterally and pre-emptively set the values of decision variables under his control in an attempt to optimize his objective function within the over-all constraint set and in cognizance of the consequent actions of lower level decision makers. We shall call such problems pre-emptive hierarchical programming, PHP, problems.

Evidently, the members of the hierarchy practice user optimization rather than system optimization. As a result the solution to the PHP problem can exhibit gross system suboptimality in the sense that there could be another solution that satisfies all the constraints and yields a higher pay-off for each of the decision makers and yet that point is unattainable (Bard, 1983 and Nwosu, 1983).

In this class of problems, the series resource control two-level PHP problem with linear constraints and objective functions has attracted much attentiion in the literature. The qualifier "series" means that there is one decision maker at each level. A number of algorithms have been proposed to solve this problem. Assuming that the constraint space S for the problem is bounded and that a unique solution exists for the lower level decision maker for any feasible choice of the upper level decision maker, Bialas and Karwan (1981) have shown that the solution of the problem must occur at an extreme point of S. Based on this result, they proposed the "K-th Best" algorithm that finds the optimal solution by an explicit (partial) enumeration of the extreme points of S.

In the Kuhn-Tucker approach, the problem of the lower level decision maker is replaced by its Kuhn-Tucker conditions. Fortuny-Amat and McCarl (1981) enforced the complementarity conditions by embedding them in a much larger mixed-integer programming problem. On the other hand, Bard and Falk (1982) solved the resultant problem using a non-convex programming algorithm based on a branch and bound technique. The Kuhn-Tucker approach has also been used by Bialas, Karwan and Shaw (1980) in the development of their parametric complementary pivot, PCP, algorithm. Their algorithm parametrically increases the value of the objective function of the upper level decision maker. Wen (1981) proposed a revised PCP algorithm to solve the problem. Candler and Townsley (1982) have proposed an implicit search scheme that generates global information at each iteration. Recently, Narula and Nwosu (1983) proposed a branch and bound procedure and Bard (1983) described a grid search algorithm, GSA, to solve the problem. From the computational results reported by Bard (1983), the GSA appears to be the most efficient algorithm at present.

In this paper, we present an efficient algorithm to solve the series resource control two-level PHP problem using the regular simplex algorithm with a few modifications. The rest of the paper is organized as follows: In Section 2 we give the mathematical programming formulation of the problem. In Section 3 we state a few observations useful in the development of the algorithm which is given in Section 4. We illustrate the proposed algorithm with an example in Section 5, and conclude the paper with a summary in Section 6.

2. PROBLEM FORMULATION

Consider an organization that has two decision making levels with one decision maker at each level, i.e., an organization with a series two-level hierarchy. Let level 1 be the higher level and DMi denote the decision maker at level i, i = 1,2, Suppose that the n decision variables are partitioned between the two decision makers such that DM1 controls $x' = (x_1, x_2, \ldots, x_{n_1})$ and DM2 controls $y' = (y_1, y_2, \ldots, y_{n_2})$ where $n_1 + n_2 = n$. Let DMi maximize the function $f_i: R^n \text{--} R^1$ for i = 1, 2. We use the following standard notation,

$$\max\{g(a, b) : (a|b)\} \equiv \max_a \{g(a, b)\} \text{ for fixed } b,$$

in which (a|b) is replaced by (a) when the value of b is not pre-set. The series two-level sequential decentralized decision making problem can be formulated as the following two-level PHP problem:

(PHP)

$$\max \{f_1(x, y) : (x)\}$$

$$\text{P1} \left\{ \begin{array}{l} \text{St.} \\ \text{P2} \end{array} \left\{ \begin{array}{l} \max \{f_2(x, y) : (y|x)\} \\ \qquad \text{St.} \quad (x, y) \in S \end{array} \right. \right.$$

where S denotes the constraint region and Pi denotes the problem to be solved by DMi, i = 1,2. The vector (x^*, y^*) is a solution of PHP if y^* solves P2 with $x = x^*$, and (x^*, y^*) solves P1.

For convenience we follow the established practice of writing PHP as follows:

$$\max \{f_1(x, y): (x)\}$$

$$\max \{f_2(x, y): (y|x)\}$$

Subject to

$$(x, y) \in S.$$

If f_i, i = 1,2 and S are all linear, we obtain the resource control PHP problem, RCP:

(RCP) $$\max \{f_1(x, y) = a_1' x + a_2' y : (x)\}, \tag{1.1}$$

$$\max \{f_2(x, y) = d_1' x + d_2' y : (y|x)\}, \tag{1.2}$$

Subject to

$$A_1 x + A_2 y \leq c, \tag{1.3}$$

$$x \geq 0, \ y \geq 0, \tag{1.4}$$

where A_i is a m x n_i matrix, i = 1,2 and c is a m x 1 vector. The lower level decision maker, DM2 solves the following problem:

(PL2) $$\max_y \{f_2(x, y) = d_1' x + d_2' y\} \tag{2.1}$$

Subject to

$$A_2 y \leq c - A_1 x, \tag{2.2}$$

$$y \geq 0, \tag{2.3}$$

for a given x.

Let S(x) be the solution set of PL2 for each fixed x. Then $\bar{S} = \bigcup_{x \varepsilon S} S(x)$ is the set of optimal responses of DM2 to the action of DM1.

Now the problem of DM1 or equivalently RCP may be written as PL1:

$$\text{(PL1)} \qquad \max \{f_1(x, y) = a_1'x + a_2'y\}$$

$$\text{Subject to}$$

$$(x, y) \varepsilon \bar{S}.$$

Although PL2 is a linear programming problem, PL1 is not since $\bar{S}$ is only piece-wise linear and hence non-convex, Bialas and Karwan (1982) and Nwosu (1983). The problems PL1 and PL2 are often referred to as the policy and behavioral problems, respectively.

To solve RCP, we break up the problem into two parts. We first generate $\bar{S}$ the optimal response set of the lower level decision maker DM2 for all values of x. Then we seek $(x^*, y^*) \varepsilon \bar{S}$ which maximizes $f_1(x, y)$.

3. SOME OBSERVATIONS

In this section, we introduce some notation and make some preliminary observations before stating the algorithm in the next section.

3.1 Notation and Preliminary Results

Our objective is to solve RCP - the problem of the upper level decision maker. This can be accomplished by first solving PL2 and then PL1. Instead of solving PL2 and PL1, we propose to solve their duals.

The dual DL2 of PL2 is:

(DL2) $$\min\ d_1' x + (c - A_1 x)' u$$

Subject to

$$-A_2' u \leq -d_2,$$

$$u \geq 0,$$

where $u \in R^m$ is the dual variable. The initial standard tableau T_{00} for DL2 can be written as

T_{00}:

1	m	n_2	
$-d_1' x$	$(c - A_1 x)'$	$0'$	1
$-d_2$	$-A_2'$	I	n_2

In our solution procedure, we shall need an updated value of $f_1(x, y)$ for DL1; therefore we append $-f_1(x, y)$ as the first column of T_{00} to obtain T_0. For convenience and clarity of exposition, we introduce the following convention in writing T_0 and subsequent tableaux. We shall expand the objective function row of T_0 to $n_1 + 1$ rows and enter only the constant term and coefficient of x in the tableaux. That is, in each column, row i ($i = 0,1,\ldots,n_1$) will contain the coefficient of x_i ($i \geq 1$) or the constant term ($i = 0$). Also to the left of the first column of the tableau, we shall write 1 in row 0 and x_i in row i ($i = 1,\ldots,n_1$) to denote the variable corresponding to row i. With this convention, T_0 can be written as:

T_0:

	$-f_1$	$-f_2$				
1	0	0	c'	$0'$	1	Objective Function
x	$-a_1$	$-d_1$	$-A_1'$	0	n_1	Rows
	$-a_2$	$-d_2$	$-A_2'$	I	n_2	Constraints Rows
	1	1	m	n_2		

Starting with T_0 we solve DL2 by using regular simplex iterations in an attempt to make $-d_2 \geq 0$. An optimal tableau T_* can be written as

T_*:

1	a_0^*	d_0^*	$c^{*\prime}$	$h^{*\prime}$	1
x	a_1^*	d_1^*	$A_1^{*\prime}$	$H_1^{*\prime}$	n_1
	a_2^*	d_2^*	$A_2^{*\prime}$	$H_2^{*\prime}$	n_2
	1	1	m	n_2	

where $d_2^* \geq 0$. Note that if it is not possible to make d_2^* greater than or equal to zero, then f_2 (x, y) is unbounded in y and the algorithm terminates at this point. The tableau T_* represents a point $(x, y) \varepsilon \bar{S}$ and it gives the optimal response of DM2 -- $y = h^* + H_1^* x$ (which is linear in x) -- to a choice of x by DM1. We need to generate other points in $\bar{S}$. To accomplish this we perform regular simplex iteration on T_* such that after each iteration the basic characteristic of T_* is maintained, i.e., $d_2^* \geq 0$. In other words, we continue to identify the outgoing basic variable by the minimum ratio rule in d_2^*. We call such pivots optimal behavioral, OB, pivots.

Definition 3.1: An optimal behavioral (OB) pivot on a non-basic column of T_* is a regular simplex pivot in A_2^* or H_2^* that uses the minimum ratio rule in d_2^* (column 2 of T_*) of the lower level objective function to identify the variable leaving the basis.

The proposed algorithm will use OB pivots to search for and locate the point in $\bar{S}$ that maximizes $f_1(x, y)$. Three tests, viz., negative coefficient test, cost coefficient test and scalar product test (see Phase III of the algorithm for details) are used to identify the incoming variable. Of the three tests, negative coefficient test is applied most frequently. Besides these three rules, the following column selection priority rule is always applied.

Column Selection Priority Rule: In selecting the next column ($j \geq 3$) to be made basic during an OB pivot in T_* the "constraint" columns, $3 \leq j \leq m+2$, must be selected before the "response" or slack columns, $m + 3 \leq j \leq m + n_2 + 2$. That is, a "response" column can only be a pivot column if no "constraint" column satisfies the applicable column selection test.

3.2 Explicit evaluation of the Optimal Point

We know that a tableau of the form T_* represents an extreme point $(x, y) \in \bar{S} \subseteq S$. But, unlike in ordinary linear programming problems we cannot immediately read off the explicit value of (x, y). However, at the current extreme point, the problem PL1 can be written as

$$\max \quad \{f_1(x, y) = -a_0^* - a_1^{*\prime}x\} \tag{3.1}$$

Subject to

$$c^* + A_1^*x \geq 0, \tag{3.2}$$

$$y = h^* + H_1^*x \geq 0, \tag{3.3}$$

$$x \geq 0. \tag{3.4}$$

This is a linear programming problem and its dual DL1 is

$$\text{(DL1)} \qquad \min \quad - a_0^* + (c^{*\prime}, h^{*\prime})\, v \tag{4.1}$$

Subject to

$$(A_1^{*\prime}, H_1^{*\prime})\, v \leq a_1^*, \tag{4.2}$$

$$v \geq 0, \tag{4.3}$$

with an initial standard tableau T_1:

$$T_1: \quad \begin{array}{|c|c|c|c|c} \hline a_0^* & c^{*\prime} & h^{*\prime} & 0^\prime & 1 \\ \hline a_1^* & A_1^{*\prime} & H_1^{*\prime} & I & n_1 \\ \hline 1 & \multicolumn{2}{c|}{t} & n_1 & \end{array}$$

Note T_1 is easily obtained from T_*. Any column of zeros in T_1 is deleted and we shall always assume that this has been done so that $t \geq m$ and not equal to $m + n_2$. The solution of T_1 yields explicit values of x and hence of (x, y) as stated in the next theorem.

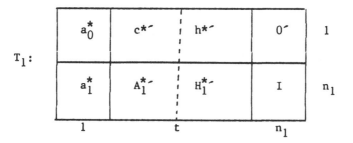

Theorem 3.1: At any current point represented (implicitly) by T_*, let the optimal tableau corresponding to its T_1 tableau be T_{1*}. Then the point (x, y) in T_{1*} is specified as follows:

(i) the vector of x is the vector of reduced cost coefficients of T_{1*} corresponding to the slack variables of T_1;

(ii) consider $(h^*, H_1^*)'$, the n_2 columns of slack variables in the objective function rows of T_*.

(a) If a column j is equal to zero then $y_j = 0$;

(b) if a column j is not equal to zero, then y_j is given by the reduced cost coefficient of T_{1*} corresponding to h_j^* of T_1.

Proof: (i) The dual of v in DL1 is x in PL1. Hence the reduced cost coefficients of the slack variables of DL1 give x.

(ii) (a) Since $y = h^* + H_1^* x \geq 0$; $(h_1^*, H_1^*)_j' = 0$ implies that $y_j = 0$.

(b) For those $y_j \neq 0$, as per (ii)(a), $y = h^* + H_1^* x \geq 0$ implies that $-H_1^* x + y = h^*$, $y \geq 0$.

Therefore y is the slack variable in the primal PL1. Hence its value given by the reduced cost coefficients of the corresponding columns in the optimal dual tableau, namely T_{1*} .

4. ALGORITHM

We shall describe the algorithm in four phases. In Phase I we set up the problem as in tableau T_0. Phase II finds an optimal solution to the dual of the behavioral problem DL2. Phase III checks and finds the global optimal solution to the dual of the policy problem DL1 in implicit form. In Phase IV we obtain the explicit solution of DL1. The details of the algorithm are as follows:

Phase I: Initialize.

Write down the tableau T_0 and go to Phase II.

Phase II: Find Optimal Solution of Behavioral Problem.

Step 1: Apply regular simplex pivots in T_0 to make all entries of $-d_2$ non-negative. Relabel final tableau T_* and go to Step 2.

Step 2: If all entries of d_2^* of tableau T_* are non-negative, go to Phase III; otherwise go to Step 3.

Step 3: STOP; $f_2(x, y)$ is unbounded in y.

Phase III: Find Global Optimal Solution to the Policy Problem:

This phase starts with the final tableau T_* of Phase II. In this phase the column selection priority rule (Section 3.1) is always applied. To state other rules that identify the incoming variable we need the following definitions. Let $p_j' = (p_{0j}, p_{1j},\ldots,p_{n_1j})$ denote the column of objective function entries of column j, $j=1,\ldots,m + n_2 + 2$. Further let

$I = \{i | i\varepsilon\{1,2,\ldots,n_1\}, a_{1i}^* < 0\}$, where $a_1^* = (a_{11}^*, a_{12}^*,\ldots,a_{1n_1}^*)'$, then ξ_j, $(3 \leq j \leq m+n_2+2)$, the coefficients of a surrogate objective function g(v,w), where w is the vector of slack variables for DL1 are given by $\xi_j = \sum_{r\varepsilon I} p_{rj}$. Note ξ_j is the sum of the coefficients of x_r, $r \varepsilon I$ in column j.

We also define the following set of integers:

$J_- = \{j | j \geq 3$, column j is an OB pivot column of T_*, $\xi_j \leq 0\}$

$J_+ =$ $\{j \mid j \geq 3$, column j is an OB pivot column of T_*, $p_1' p_j < 0\}$

$H = \{j \mid j \geq 3$, column j is an OB pivot column of T_*, the constant row (row 0) entry of column j is negative; <u>after the proposed pivot</u> the new $a_1^* \geq 0$ $\}$

$F =$ $\{j \mid j \geq 3$, $p_j \neq 0$ and $p_j \leq 0\}$

$G = \{j \mid j \in F$ and column j is an OB pivot column of $T_*\}$

<u>Step 1</u>: (Negative Coefficient Test)

Evaluate J_-. If $J_- \neq \phi$, select j, $j \in J_-$, go to Step 2; otherwise, go to Step 3.

<u>Step 2</u>: Perform an OB pivot in column j of the current tableau T_*. Go to Step 1.

<u>Step 3</u>: If $I \neq \phi$ (i.e., $a_{1i}^* < 0$ for some $1 \leq i \leq n_1$), go to Step 5; otherwise, go to Step 4A.

<u>Step 4A</u>: If $c^* \geq 0$ and $h^* \geq 0$, go to Step 4C; otherwise go to Step 4B.

<u>Step 4B</u>: (Cost Coefficient Test)

Evaluate H. If $H = \phi$, go to Step 5; otherwise select j, $j \in H$, and go to Step 2.

<u>Step 4C</u>: (Connectedness Test)

If $a_2^* \geq 0$, go to Step 5; otherwise go to Step 4D.

<u>Step 4D</u> (Scalar Product Test)

Evaluate J_+. If $J_+ \neq \phi$, select j, $j \in J_+$, and go to Step 2; otherwise go to Step 5.

<u>Step 5</u>: If $F = \phi$ go to Phase IV; otherwise go to Step 6.

Step 6: (Primal Feasibility Test)

If $G = \phi$ go to Step 7; otherwise select j, $j \in G$, and go to Step 2.

Step 7: STOP. Problem is infeasible.

Phase IV: Find Explicit Solution of RCP.

Step 1: Write down T_1 from the final tableau of Phase III.

Step 2: Solve the linear programming problem T_1. Read (x, y). STOP.

5. AN EXAMPLE

We illustrate the proposed OB pivot algorithm with the following example from Bard (1983).

$$\max \{2x_1 - x_2 - x_3 + 2x_4 + x_5 - 3.5x_6 - y_1 - 1.5y_2 + 3y_3 : (x)\}$$

$$\max \{ 2x_2 - x_5 + 3y_1 - y_2 - 4y_3 : (y|x)\}$$

Subject to

$$\begin{aligned}
-x_1 + .2x_2 + x_5 + 2x_6 - 4y_1 + 2y_2 + y_3 &\leq 12 \\
x_1 + x_3 - 2x_4 - 4y_2 + y_3 &\leq 10 \\
5x_1 + x_4 + 3.2x_6 + 2y_1 + 2y_2 &\leq 15 \\
- 3x_2 - x_4 + x_5 - 2y_1 &\leq 12 \\
- 2x_1 - x_2 - y_2 + y_3 &\leq -2 \\
- y_1 - 2y_2 - y_3 &\leq -2 \\
- 2x_2 - 3x_3 - x_5 &\leq -3
\end{aligned}$$

where $x_i \geq 0$, $i = 1, \ldots, 6$ and $y_i \geq 0$, $i = 1,2,3$.

The initial tableau T_0 for the problem is given in Table 1.

TABLE 1

Initial Tableau T_0 for the Example

	$-f_1$	$-f_2$										
1	0	0	12	10	15	12	−2	−2	−3	0	0	0
x_1	−2	0	1	−1	−5	0	2	0	0	0	0	0
x_2	1	−2	−1/5	0	0	3	1	0	2	0	0	0
x_3	1	0	0	−1	0	0	0	0	3	0	0	0
x_4	−2	0	0	2	−1	1	0	0	0	0	0	0
x_5	−1	1	−1	0	0	−1	0	0	1	0	0	0
x_6	7/2	0	−2	0	−16/5	0	0	0	0	0	0	0
	1	−3	4	0	−2	2	0	1	0	1	0	0
	3/2	1	−2	4	−2	0	1	2	0	0	1	0
	−3	4	−1	−1	0	0	−1	1	0	0	0	1

One pivot operation (pivot element is circled in Table 1) gives us Table 2, in which all elements of $d_2^* \geq 0$. This ends Phase II of the algorithm. To execute Phase III of the algorithm, we append the column numbers (above the Table) and the coefficients of the surrogate function (below the Table) to Table 2.

TABLE 2

Tableau T_* at the End of Phase II for the Example

	1	2	3	4	5	6	7	8	9	10	11	12
1	15/2	-45/2	42	10	0	27	-2	11/2	-3	15/2	0	0
x_1	-9/2	15/2	-9	-1	0	-5	2	-5/2	0	-5/2	0	0
x_2	1	-2	-1/5	0	0	3	1	0	2	0	0	0
x_3	1	0	0	-1	0	0	0	0	3	0	0	0
x_4	-5/2	3/2	-2	2	0	0	0	-1/2	0	-1/2	0	0
x_5	-1	1	-1	0	0	-1	0	0	1	0	0	0
x_6	19/10	24/5	-42/5	0	0	-16/5	0	-8/5	0	-8/5	0	0
	-1/2	3/2	-2	0	1	-1	0	-1/2	0	-1/2	0	0
	1/2	4	-6	4	0	-2	1	1	0	-1	1	0
	-3	4	-1	-1	0	0	-1	1	0	0	0	1
			-12	1	0	-6	2	-3	1	-3	0	0

In Table 2, set I = {1, 4, 5} and set J_- = {8}, which gives us the pivot element (circled in Table 2). After the pivot operation, we obtain Table 3.

TABLE 3

Tableau T_* After the First OB Pivot

	1	2	3	4	5	6	7	8	9	10	11	12
1	19/4	-89/2	75	-12	0	38	-15/2	0	-3	13	-11/2	0
x_1	-13/4	35/2	-24	9	0	-10	9/2	0	0	-5	5/2	0
x_2	1	-2	-1/5	0	0	3	1	0	2	0	0	0
x_3	1	0	0	-1	0	0	0	0	3	0	0	0
x_4	-9/4	7/2	-5	4	0	-1	1/2	0	0	-1	1/2	0
x_5	-1	1	-1	0	0	-1	0	0	1	0	0	0
x_6	27/10	56/5	-18	32/5	0	-32/5	8/5	0	0	-16/5	8/5	0
	-1/4	7/2	-5	2	1	-2	1/2	0	0	-1	1/2	0
	1/2	4	-6	4	0	-2	1	1	0	-1	1	0
	-7/2	0	(5)	-5	0	2	-2	0	0	1	-1	1
			-30	13	0	-12	5	0	1	-6	3	0

In Table 3, set $I = \{1, 4, 5\}$, set $J_- = \{3, 6, 10\}$ and the pivot element is circled (in Table 3). The pivot operation results in Table 4.

TABLE 4

Tableau T_* After the Second OB Pivot

	1	2	3	4	5	6	7	8	9	10	11	12
1	229/4	-89/2	0	63	0	8	45/2	0	-3	-2	19/2	-15
x_1	-401/20	35/2	0	-15	0	-2/5	-51/10	0	0	-1/5	-23/10	24/5
x_2	43/50	-2	0	-1/5	0	77/25	23/25	0	2	1/25	-1/25	1/25
x_3	1	0	0	-1	0	0	0	0	3	0	0	0
x_4	-23/4	7/2	0	-1	0	1	-3/2	0	0	0	-1/2	1
x_5	-17/10	1	0	-1	0	-3/5	-2/5	0	1	1/5	-1/5	1/5
x_6	-99/10	56/5	0	-58/5	0	4/5	-28/5	0	0	2/5	-2	18/5
	-15/4	7/2	0	-3	1	0	-3/2	0	0	0	-1/2	1
	-37/4	4	0	-2	0	2/5	-7/5	1	0	1/5	-1/5	6/5
	-7/10	0	1	-1	0	2/5	-2/5	0	0	1/5	-1/5	1/6
			0	-143/5	0	4/5	-63/5	0	1	2/5	-5	43/5

In Table 4, set I = {1, 4, 5, 6} but $J_- = \phi$. At this stage, we end Phase III of the algorithm.

From Table 4, we obtain T_1 as shown in Table 5.

TABLE 5

Tableau T_1 for the Example

					y_1	y_2	y_3						
229/4	63	8	45/2	-3	-2	19/2	-15	0	0	0	0	0	0
-401/20	-15	-2/5	-51/10	0	-1/5	-23/10	24/5	1	0	0	0	0	0
43/50	-1/5	77/25	23/25	2	1/25	-1/25	1/25	0	1	0	0	0	0
1	-1	0	0	3	0	0	0	0	0	1	0	0	0
-23/4	(-1)	1	-3/2	0	0	-1/2	1	0	0	0	1	0	0
-17/10	-1	-3/5	-2/5	1	1/5	-1/5	1/5	0	0	0	0	1	0
-99/10	-58/5	4/5	-28/5	0	2/5	-2	18/5	0	0	0	0	0	1

Note that in order to keep track of the vector y in Table 5, we give the column labels for the explicitly non-zero components of y (all three in this example). Four regular simplex iterations give us the optimal tableau for the problem as shown in Table 6.

TABLE 6

Optimal Tableau for the Example

					y_1	y_2	y_3						
-206/5	38	149/5	0	71/5	0	0	2	0	4	0	15	46/5	0
23/5	-11	-62/5	0	-13/5	0	0	0	1	-2	0	-333/25	7/5	0
6/5	0	16/5	1	9/5	0	0	0	0	1	0	0	-1/5	0
1	-1	0	0	3	0	0	0	0	0	1	0	0	0
79/10	2	-58/5	0	-27/5	0	1	-2	0	-3	0	-2	3/5	0
9/5	-3	-41/5	0	16/5	1	0	-1	0	-1	0	-2	26/5	0
119/10	-32/5	-6/5	0	-2	0	0	0	0	0	0	-16/5	-2	1

The solution is $x' = (0, 4, 0, 15, 46/5, 0)$ and $y' = (0, 0, 2)$ with $f_1(x, y) = 206/5$ and $f_2(x, y) = -46/5$.

Since the last entry of column 2 in the final tableau for Phase III (Table 4) is zero (i.e., some component of a_2^* is zero), there is the possibility of the existence of alternate solution to the primal problem. On investigation we found that the solution in Table 6 is unique.

6. SUMMARY

We have presented an efficient algorithm to solve the series resource control pre-emptive hierarchical programming problem using regular simplex pivots with a few modifications to select variables to enter and leave the basis. These modifications guarantee that the proposed algorithm will find the global optimal solution to the problem, when it exists and identify cases when the problem is either infeasible or unbounded. The algorithm was developed by considering explicitly the sequential nature of the decision making process for this problem and the partition of the decision variables between the two decision makers.

REFERENCES

Bard, J. F. (1983). An efficient point algorithm for a linear two-stage optimization problem. Operations Research, 31, 670-684.

Bard, J. F. and Falk, J. E. (1982). An explicit solution to the multi-level programming problem. Comput. Opns. Res., 9, 77-100.

Bialas, W. F. and Karwan, M. H. (1981). Two-level linear programming: A primer. Technical Report, Department of Industrial Engineering, SUNY at Buffalo.

Bialas, W. F. and Karwan, M. H. (1982). On two-level optimization. IEEE Transactions on Automatic Control, AC-27, 211-214.

Bialas, W. F., Karwan, M. H. and Shaw, J. P. (1980). A parametric complementary pivot approach for two-level linear programming. Research Report No. 80-2, Dept. of Industrial Engineering, SUNY at Buffalo.

Candler, W. and Townsley, R. (1982). A linear two-level programming problem. Comput. Opns. Res., 9, 59-76.

Fortuny-Amat and McCarl, B. (1981). A representation and economic interpretation of a two-level programming problem. J. Operational Res. Soc., 32, 783-792.

Narula, S. C. and Nwosu, A. D. (1983). Two-level hierarchical programming problem. Multiple Criteria Decision Making - Theory and Application (P. Hansen, Editor), Springer Verlag, 290-299.

Nwosu, A. D. (1983). Pre-emptive hierarchical programming problem: A decentralized decision model. Unpublished Ph.D. dissertation. Department of Operations Research and Statistics, Rensselaer Polytechnic Institute, Troy, New York.

Wen, U. (1981). Mathematical methods for multi-level linear programming. Unpublished Ph.D. dissertation, Department of industrial Engineering, SUNY at Buffalo, Buffalo, New York.

COMPOSITE PROGRAMMING AS AN EXTENSION OF COMPROMISE PROGRAMMING

Andras Bárdossy, Istvan Bogárdi
Tiszadata Consulting Engineers

Lucien Duckstein
Systems and Industrial Engineering Department
University of Arizona

ABSTRACT

Cost-effective control alternatives are selected by composite programming /an extension of compromise programming/ in order to find a trade-off among objectives or groups of criteria usually facing watershed management or observation network design: economic criteria /agricultural revenue and investment/, environmental criteria /yields of sediment and nutrient/, and hydrologic criteria /water yield/. Composite programming provides a two-level trade-off analysis: first with different L_p-norms within the criteria, then again with a different L_p-norm among the three objectives.

An example of six interconnected watersheds draining

into a multipurpose /water supply and recreation/ reservoir and a network design problem of aquifer parameters illustrate the methodology.

INTRODUCTION

The purpose of this paper is to develop an extension of compromise programming, called composite programming and apply it to multicriterion watershed management, and observation network design. Agricultural watersheds often serve conflicting objectives such as the maximization of agricultural revenue, minimization of water pollution and maximization of water yield from the watershed. Also, each of these objectives may have several components; for example, the first objective may be composed of annual income and investment cost. The second objective may be characterized by the yields of sediment, nutrients, heavy metals.

Next, geostatistics and multicriterion decision making are combined to design a regular observation network for several spatially correlated and anisotropic parameters. The decision variables are: network density, distance between observation points for the various parameters and observation effort. The estimation error is calculated as a function of the decision variables by use of a geostatistical model based on prior variograms. Composite programming, an extension of compromise programming with more than one value of p in the L_p distance, makes possible to account for the analytical characteristics of statistical criteria versus the economic value of observation effort. Thus, an L_1-metric is applied to statistical criteria and an L_2-metric, to observation effort criteria. The algorithm for solution uses gradient optimization. The example of a two-layer aquifer system where thickness and porosity are the parameters to be identified illustrates the methodology. The composite solution appears to be quite robust to changes of the weights assigned to parameters.

COMPROMISE PROGRAMMING

Compromise programming /Zeleny[1], Gershon[2]/ is an approach which defines the "best" solution as that point which minimizes the distance from a goal point /often the ideal point is used/ to the set of efficient solutions. By restricting the goal point so that it is greater than or equal to the ideal point /see Figure 1/, the affect of choosing this point is minimized. The motivation behind this method is the desire to achieve a solution that is as "close" as possible to some "ideal".

The distance measure used in compromise programming is the family of L_p-metrics. This metric is given as:

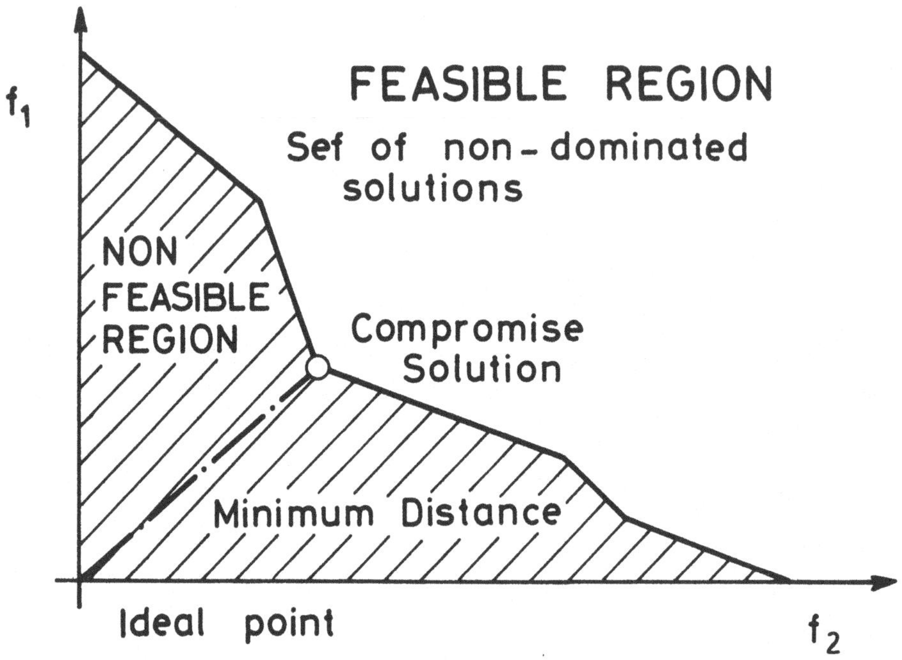

Figure 1. Selection of goal points by compromise programming

$$L_p(\underline{x}) = \left[\sum_{i=1}^{n} \alpha_i^p \left| \frac{f_i^* - f_i(\underline{x})}{f_i^* - f_{i\text{-min}}} \right|^p \right]^{1/p} \qquad /1/$$

where the α_i are the weights, f_i^* is the optimal value of the ith criterion, $f_{i\text{-min}}$ is the worst value obtainable for criterion i, and $f_i(x)$ is the result of implementing decision $\underline{x}$ with respect to the ith criteria. For p = 1, all deviations from f_i^* are taken into account in direct proportion to their magnitudes. For $2 \leq p < \infty$, the largest deviation has the greatest influence. For $p = \infty$, the largest deviation is the only one taken into account /minimax criterion/.

COMPOSITE PROGRAMMING

Composite Programming has been introduced as an empirical technique to resolve a geological exploration problem in Bogárdi and Bárdossy[3]. The analysis of this technique consists of three parts. First, it is shown that the composite programming expression satisfies the mathematical properties of distances. Next, it is proved that composite programming is a real extension of compromise programming, that is, compromise distance is a special case of composite distance, but not every composite distance can be replaced by an equivalent compromise distance. Finally, it is demonstrated that every solution minimizing a composite distance also minimizes an L_2-distance with different weights.

Composite Programming as a distance

Composite programming is a distance-based technique in which the function

$$\varrho\,(X, X^*) \qquad \text{is minimized,}$$

where X represents the objective function vector corresponding to a feasible solution, and X^* is the ideal point.

The distance is a composite distance as defined below.

Definition: A distance ϱ is called composite metric on $\mathbb{R}^n$ if for each $\underline{x}$, $\underline{y} \in \mathbb{R}^n$ we have

$$\varrho/x,\ y/ = \left(\sum_j \beta_j^q \left(\sum_{i \in I_j} \alpha_i^{p_j} |x_i - y_i|^{p_j}\right)^{q/p_j}\right)^{1/q} \qquad /2/$$

where

$$\alpha_i,\ \beta_i > 0 \quad \text{and } \infty > p_i \geq 1,\ 1 \leq q < \infty \quad \bigcup_i I_i = \{1, 2, \dots, n\}$$

Theorem 1: ϱ is a metric on $\mathbb{R}^n$

Proof: One has: $\varrho/\underline{x},\underline{y}/ =$ if and only if $\underline{x} = \underline{y}$

and $\varrho/\underline{x},\underline{y}/ = \varrho/\underline{y},\underline{x}/$.

As $\varrho/\underline{x},\underline{y}/ = \varrho/\underline{x}-\underline{y},\ \underline{o}/$ we only have to show

that $\varrho/x,o/ = |||x|||$ is a norm.

This means $|||x||| + |||y||| \geq |||x+y|||$,

which can be shown by simple induction:

Let $\|.\|_1$ be a norm on $\mathbb{R}^{n1}$, $\|.\|_2$ a norm on $\mathbb{R}^{n2}$

then

$$\|\underline{x},\underline{y}\|_3 = \left((\|\underline{x}\|_1)^p + (\|\underline{y}\|_2)^p\right)^{1/p} \qquad /3/$$

is a norm on $\mathbb{R}^{n1+n2}$.

$$|||\underline{x}_1+\underline{x}_2,\ \underline{y}_1+\underline{y}_2||| = \left((\|\underline{x}_1+\underline{x}_2\|_1)^p + (\|\underline{y}_1+\underline{y}_2\|)^p\right)^{1/p} \leq$$

$$\leq \left((\|x_1\|_1 + \|x_2\|_1)^p + (\|y_1\|_2 + \|y_2\|_2)^p\right)^{1/p} \leq$$

$$\left((\|x_1\|_1)^p + (\|y_1\|_2)^p\right)^{1/p} + \left((\|x_2\|_1)^p + (\|y_2\|_2)^p\right)^{1/p} = \qquad /4/$$

$$= |||\underline{x}_1,\ \underline{y}_1||| + |||\underline{x}_2,\ \underline{y}_2|||$$

Composite Programming as an extension of compromise programming

The following example shows that composite programming is a proper extension of compromise programming, that is, a composite metric cannot be replaced by a compromise metric with similar weighting.

Example: $/x_1+x_2/^2 + x_3{}^2 \geq \underline{1} \qquad x_i \geq 0$ /5/

$$x_1^P + x_2^P + \alpha\, x_3^P \longrightarrow \min \qquad /6/$$

If a point $/x_1^+,\ x_2^+,\ x_3^+/$ is stationary, then we have $x_1^+=x_2^+$ because, by the Kuhn-Tucker conditions:

$$P \cdot x_1^{P-1} - \lambda_1\, 2/x_1 + x_2/ = 0 \qquad /7/$$

$$P\, x_2^{P-1} - \lambda_1\, 2\, /x_1 + x_2/ = 0 \qquad /8/$$

$$P \cdot \alpha \cdot x_3^{P-1} - \lambda_1\, 2 \quad x_3 = 0 \qquad /9/$$

$$\lambda_1 \left(/x_1 + x_2/^2 + x_3^2 - 1 \right) = 0 \qquad /10/$$

This means that there are points which are optimal with respect to a composite distance but are non-optimal with respect to a compromise distance with a similar structure of weights.

Relationship between composite distance and L_2-distance

The following theorem shows that the nondominated points which correspond to the minima of composite distances are also minima for compromise distances - namely a weighted L_2-distance.

Theorem 2:

Let $/x_1^*, \dots, x_n^*/ \ \mathbb{R}_+^n$ be such that $\varrho\,/\underline{x}^*, \underline{o}/ = \underline{1}$ for some composite distances. Then there are weights $\gamma_i \geq 0$ such that

$$\sum_{i=1}^{n} \gamma_i x_i^2 \geq \sum_{j=1}^{n} \gamma_i x_i^{*2} \qquad /11/$$

if $\varrho\,/\underline{x}, \underline{o}/ = \underline{1}, \ x \in \mathbb{R}^n$

Proof: The proof, based on the Kuhn-Tucker conditions can be found in Bárdossy[18]. Theorem 2 may be used in those frequent numerical cases when it is easier to deal with a quadratic objective function $/L_2$-distance/ than a composite distance. So we define a composite metric search on L_2 metric which minimizes the distance with respect to the L_2, then recalculate the weights of the corresponding composite metric, select a new weighted L_2, and so on.

APPLICATION EXAMPLES

Multicriterion watershed management

Three conflicting objectives of watershed management are distinguished:

i./ to increase crop yield, and thus agricultural benefit by increasing effective precipitation, that is, infiltration, and thus decreasing runoff,

ii./ to decrease water pollution stemming from agricultural activity,

iii./ to increase water yield, that is, surface and subsurface runoff in order to meet downstream water demand.

Table 1 shows that the economics related criterion /i/, and the water pollution criterion /ii/ have several components. More generally, it may also be necessary to consider components such as nitrogen and heavy metals.

While economic criteria can certainly be expressed a single monetary term such as discounted net benefit, it is often advisable to keep each criterion in a separate account; for example, agricultural revenue, investments and losses may correspond to different cost and revenue sharing schemes.

A watershed consisting of several interconnected subwatersheds is investigated. Topographical, soil and agri-

Table 1

Objectives of watershed management

Objective	Elements
Economic	Annual agricultural revenue Investment cost Cost-benefit ratio
Environmental: Water pollution	Sediment yield Dissolved phosphorus Sediment-bound phosphorus
Hydrologic: Water yield	Annual runoff

cultural conditions differ among subwatersheds. Elements of the three sets of criteria for each subwatershed /except cost-benefit ratio/ can be added to represent the whole watershed.

Rainfall as stochastic uncontrollable input influences the three criteria. As a result, most of the criterion elements should also be regarded as random variables. In the present case, the water pollution criterion may reflect the existence of a downstream water body /lake or reservoir/. In this respect, sediment yield indicates not only upstream surface erosion and/or local scour but it is adverse in view of the life-time and operation of the water-body. Dissolved phosphorus /P/ has often turned to be the triggering effect of eutrophication, and labile phosphorus bound to sediment may also become available for algea growth from reservoir deposition.

There is a crucial conflict in watershed management between the interests of watershed agriculture and downstream water users. The former endeavours to increase crop production by increasing the efficiency of rainfall utilization over the watershed.

On the other hand, downstream water users such as the operators of downstream reservoirs would prefer to increase or at least to maintain runoff to meet demands of water supply, irrigation, power generation, etc. This conflict situation can be strikingly illustrated by the existing two meanings of effective precipitation. Agricultural managers generally consider as effective precipitation the ratio which infiltrates and is utilized by the vegetation, while hydrologists would call the runoff ratio as effective precipitation.

Several applications of multicriterion decision making /MCDM/ to model and resolve this type of problem can be found /Miller and Byers[4]; Das and Haimes[5]; Bogárdi et al.[6]/. Here, composite programming is applied to a two-level tradeoff analysis. At the first level, a compromise is sought within each objective, for example between sediment yield and nutrient yield within the water pollution objective. At the second level, the three objectives; economics, water pollution and water yield are traded off.

For the management of the watershed, the following goals are specified:

$$\begin{aligned} &\min\ E(\tilde{Z}) \\ &\max\ E(\tilde{V}) \\ &\min\ E(\tilde{SP}) \\ &\min\ E(\tilde{KP}) \\ &\max\ E(\tilde{H}) - K \\ &\min\ B \end{aligned} \qquad /12/$$

where ~ denotes a random variable.

$\tilde{Z}$ = annual sediment yield from the whole watershed,

$\tilde{V}$ = annual water runoff,

$\tilde{SP}$ = annual soluble phosphorus /P/ yield,

$\tilde{KP}$ = annual labile P yield,

$\tilde{H}$ = annual crop revenue,

K = annual cost of farming,

B = investment costs.

Since the whole watershed is divided into N subwatersheds, the expected values in Eq. /12/ have to be calculated for each of the alternatives over every subwatershed.

The annual sediment yield $\tilde{Z}$ is estimated by an event-based version of the modified Universal Soil Loss Equation

/MUSLE/ /Williams[7]/. This method uses rainfall events /storm depth, duration/ as input and provides fairly occurate results /Bárdossy et al.[8]/.

The annual water yield $\underset{\sim}{V}$ is calculated by the SCS method /Soil Conservation Service[9]/.

The annual amount of dissolved P /$\underset{\sim}{SP}$/ is estimated as the annual sum of a random number of random loading events as characterized by the product of runoff volume per event and the dissolved P concentration. The latter concentration can be estimated by a physico-chemical model /Bogárdi and Bárdossy[10]/.

The annual amount of soil-bound available P /$\underset{\sim}{KP}$/ is calculated similarly to $\underset{\sim}{SP}$. However, loading events are here characterized by the product of sediment yield per event and labile P concentration.

The latter quantity can be obtained in conjunction with the dissolved P concentration /Bogárdi and Bárdossy[10]/.

The annual crop revenue $\underset{\sim}{H}$ is calculated as the product of crop yield and price.

Let a /0-1/ variable k_{ji} be introduced as follows:

$$k_{ji} = \begin{cases} 1 & \text{if alternative } j \text{ is selected for subwatershed } i \\ 0 & \text{otherwise} \end{cases}$$

Since only one alternative can be selected for each subwatershed, one has:

$$\sum_{j=1}^{N_i} k_{ji} = \underline{1} \quad /\text{for } i = 1, \ldots, N/ \qquad /13/$$

Now the required expected values, corresponding to a selected set of alternatives

$$\underline{\underline{k}} = /k_{ji}/ \begin{matrix} N m_i \\ i=1, j=1 \end{matrix}$$

can be calculated.

To find optimal alternatives we select an ideal point and then we define a composite metric. The optimal solution minimizes the distance between the ideal point $\underline{y}^*$ and the point $\underline{y}$ in the pay-off space. The ideal point is:

$$\underline{y}^* = /Z_*, V^*, SP_*, KP_*, HK^*, B_*/.$$

The objective is to minimize the distance $\varrho/\underline{y}, \underline{y}^*/$. This distance is defined stepwise as stated in the composite programming formulation. First, three groups of criteria, each group corresponding to an objective, are formed. In each group a compromise solution, and then the overall compromise are sought. The groups are as follows:

- HK, B
- Z, SP, KP
- V

In each group, one selects a value of compromise programming parameter p and a set of weights $\{\alpha\}$. Then the overall objective function with parameter q and weight set $\{\beta\}$ is defined. This procedure yields the objective functions:

$$CF_1/\underline{\underline{k}}/ = \left[\alpha_{11}^{p_1}\left(\frac{HK^* - HK/\underline{\underline{k}}/}{HK^* - HK_*}\right)^{p_1} + \alpha_{12}^{p_1}\left(\frac{B^* - B/\underline{\underline{k}}/}{B^* - B_*}\right)^{p_1}\right]^{1/p_1} \qquad /14/$$

$$CF_2/\underline{\underline{k}}/ = \left[\alpha_{21}^{p_2}\left(\frac{Z/\underline{\underline{k}}/ - Z_*}{Z^* - Z_*}\right)^{p_2} + \alpha_{22}^{p_2}\left(\frac{SP/\underline{\underline{k}}/ - SP_*}{SP^* - SP_*}\right)^{p_2} + \right.$$

$$\left. + \alpha_{23}^{p_2}\left(\frac{KP/\underline{\underline{k}}/ - KP_*}{KP^* - KP_*}\right)^{p_2}\right]^{1/p_2} \qquad /15/$$

$$CF_3/\underline{\underline{k}}/ = \frac{V^* - V/\underline{\underline{k}}/}{V^* - V_*} \qquad /16/$$

The overall composite goal function is written as:

$$CF/\underline{\underline{k}}/ = \left(\beta_1^q \cdot CF_1/\underline{\underline{k}}/^q + \beta_2^q \cdot CF_2/\underline{\underline{k}}/^q + \beta_3^q \cdot CF_3/\underline{\underline{k}}/^q\right)^{1/q} \qquad /17/$$

The Vácszentlászló watershed of 27 km^2 in Hungary, which is used to illustrate the methodology, is divided into six subwatersheds with basic characteristics given in Table 2.

A storage reservoir of 640 x 10^3 m^3 serving irrigation, water-supply and offering recreational possibilities is located at the outlet of the watershed. The conflicting interests facing the management of this watershed can be illus-

Table 2

Basic data for the Vácszentlászló watershed

Sub-watershed	Area Ha	Avr. slope %	MUSLE parameters av. L	av. K	Area Ha	Avr. slope %	MUSLE parameters av. L	av. K
I	176.7	16.0	3.3	0.36	-	-	-	-
II	330.6	12.5	3.5	0.32	41.0	14.0	3.6	0.30
III	202.5	8.5	3.1	0.34	-	-	-	-
IV	246.8	11.2	3.3	0.40	-	-	-	-
V	208.8	5.8	3.1	0.36	267.1	10.0	2.9	0.34
VI	-	-	-	-	767.9	8.0	2.8	0.30
Total	1165.4				1076.0			

trated by the following fact: Under existing conditions the average annual sediment yield of 2700 tons stemming mostly from surface erosion adversely affects agricultural production, gradually decreases the storage volume of the reservoir, and contributes to the P loading. Both dissolved and sediment-bound P loadings into the reservoir cause increasing eutrophication, endangering its recreational function.

Intensive agriculture over the watershed has resulted in an average annual net agricultural revenue of $11,598\text{x}10^3$ Ft /1 dollar equals to about 50 Forints/. It is an important regional goal to maintain or possibly increase this revenue.

Concerning the third objective /hydrology/, water-supply requirements from the reservoir have been growing. Thus one seeks to increase the annual water yield from the watershed.

In order to find a proper trade-off among these objectives, a number of discrete alternative interventions over each subwatershed has been envisaged. Table 3 indicates the number of alternatives of each type, leading to a total number of 46,268,820. The use of total enumeration is thus outruled.

Table 3

Basic alternatives for the Vácszentlászló multicriterion watershed management

Type of alternatives	Number of alternatives for subwatersheds					
	I	II	III	IV	V	VI
Land use	6	6	5	5	4	1
Crop	2	2	2	2	2	1
Cultivation	3	3	3	3	3	1
Slope-length decrease	4	4	4	4	4	1
Amount of fertilizer /existing and 50 % decrease/	2	2	2	2	2	1

The MCDM model has been applied to this case with the following encoding of watershed characteristics:
each variable k_{ij} is characterized by a six-digit code wherein

- the first digit indicates the number of subwatersheds: 1, ..., 6
- the second digit shows land use:

1: existing

2: total afforestation

3: total meadow

4: cropland, on slopes > 25 % forest

5: cropland < 17 %, 17-25 % meadow, forest > 25 %

6: cropland, > 25 % meadow

- the third digit refers to crop rotation:

0: existing

1: soil protective

- the fourth digit indicates the mode of cultivation

0: up-and-down slope culture

1: contour farming

2: contour farming and crop-residue

- the fifth digit refers to erosion control by slope length control

0: no control

1: slope length: 400-1000 m
2: slope length: 250- 700 m } depending on steepness
3: slope length: 100- 500 m

- the sixth digit refers to fertilizer control

1: existing amount of P

2: 50 % decrease.

Composite programming parameters p and q are:

$p_1 = 2$; for trade-off between agricultural revenue and investment,

p_2 = 3; for trade-off among environmental elements, to emphasize the limiting character of the worst element,

q = 2; for trade-off among the three objectives.

Numerical results are given in Table 4 for several sets of weights, α and β.

Since subwatershed 6 is total forest, it will be kept intact in every solution. Sets of weights I and II correspond to a balanced importance of objectives; as a result, the composite solutions are quite similar, namely:

- the same land use changes over the subwatersheds /1-5/,
- use of protective crop rotation everywhere /1-5/,
- contour farming and plant residues everywhere /0-5/,
- slight /1/ to medium /2/ erosion control,
- existing amount of P-based fertilizer application.

Sets III and IV gives high preference to the hydrologic and environmental objectives, respectively. With highly preferred runoff maximization:

III, the composite solution changes considerably:

- maintenance of existing situation over subwatersheds 1, 3 and 6,
- decrease of fertilizer use over subwatersheds 1 and 3.

On the other hand, with a high preference for the environment /IV/ the solution corresponds to uniform land use /4/ and erosion control /medium/ over each subwatershed except No. 6.

Table 4

Results of composite programming

Objective	Element	Set of weights I: $\beta = 0.4; 0.3; 0.3$, $\alpha_1 = 0.9; 0.1$, $\alpha_3 = 0.3; 0.4; 0.3$	Set of weights II: $\beta = 0.2; 0.6; 0.2$, $\alpha_1 = 0.9; 0.1$, $\alpha_3 = 0.4; 0.3; 0.3$
Economic	Revenue 10^3 Ft/y	15762	16494
	Investment 10^3 Ft	11508	13280
Environmental	Sediment, tons	547	496
	Dissolved P	5.7	5.4
	Sed. P	5.8	5.2
Hydrologic	Runoff 10^3 m^3	105	102
Composite solution		161211	161221
		261211	261211
		341211	341211
		441211	441221
		541211	541221
		610001	610001

Table 4 cont.

Objective	Element	Set of weights III: $\beta = 0.1; 0.1; 0.8$; $\alpha_1 = 0.9; 0.1$; $\alpha_3 = 0.3; 0.2; 0.5$	Set of weights IV: $\beta = 0.1; 0.8; 0.1$; $\alpha_1 = 0.9; 0.1$; $\alpha_3 = 0.8; 0.1; 0.1$
Economic	Revenue 10^3 Ft/y	12512	17219
	Investment 10^3 Ft	7662	18021
Environ-mental	Sediment, tons	1412	415
	Dissolved P	6.7	4.7
	Sed. P	14.2	4.2
Hydrologic	Runoff 10^3 m^3	112	96
Composite solution		110002	141221
		261211	241221
		310002	341221
		441211	441221
		541211	541221
		610001	610001

Observation network design

Consider a multilayer aquifer system such as the six aquifers underlying the Venetian lagoon /Volpi et al.[11]/. Various properties of such an aquifer should be observed in order to predict parameter values at ungaged points or else, areal average values. Properties such as thickness, porosity, transmissivity are then used for calculating the underground flow or the slope stability characteristics of a pit. On the other hand, piezometric levels can also be observed in order to construct, say, groundwater maps.

Many similar observation problems can be found in water resources management. A typical example is the planning of hydrometeorological stations for observing quantities such as rainfall, temperature, radiation, soil moisture and pH value. All such parameters to be observed exhibit "geostatistical" properties in the sense of Matheron[12]; that is, they can be assumed to have a deterministic component plus a spatially correlated and anisotropic stochastic component. The observation network design itself may involve the following interrelated factors:

i/ observation effort /cost, time, instrumentation, etc./,

ii/ relative importance of the various parameters,

iii/ the different geostatistical properties of the parameters, iv/ estimation accuracy or error criteria for the various parameters.

Thus, two sets of objectives may be considered:

1/ to minimize the criteria related to estimation error,

2/ to minimize the criteria related to the observation effort with due regard to the various importance and geostatistical properties of the parameters.

The parameters to be observed generally change in space and time. The variability in time can be short-term, such as temperature or rainfall, medium-term such as groundwater depth or top soil pH value, or long-term such as the formation of geologic layers. In this paper, only simultaneous values of the parameters are considered, that is, only spatial behavior is to be observed. Consider, thus, an area where an exploration network is being planned to measure the thickness and porosity of a two-layer aquifer. Let this network be characterized by the density of stations /m^{-2}/ or its reciprocal, T.

The following two problems, whose solutions are functions of observation stations, are considered:

i/ predict parameter values from measurement data at non-explored points, and

ii/ estimate averages for areas, blocks or volumes.

The network can be represented by three decision variables:

T: area covered by one station, that is, the reciprocal of the density of stations,

a: average distance between two stations measured along network orientation,

Θ: angle between network orientation and N-S direction.

The efficiency of estimation can be characterized by the estimation variances. For parameter k, this variance σ^2_k is a function of the decision variables T, a, and Θ. Thus, one set of objectives refers to

$$\min \sigma_k /T, a, \Theta/ \qquad k = 1, \ldots, K \qquad /18/$$

where K is the number of variables to be observed.

A geostatistical procedure has been developed to calculate function /18/ /Bárdossy and Bogárdi[3]/.

Another set of objectives refers to minimization of the observation effort characterized, here, by a single criterion, the observation cost, which increases with the density of stations T^{-1}. One thus seeks to minimize T^{-1} /or maximize T/.

Such an observation network is sought which is the closest by some distance measure, or metric, to the ideal one. To define this metric, consider the family of L_p metrics used in compromise programming /Equation /1/ /.

This L_p metric is modified to account for the different nature of the two sets of criteria /statistical and observation effort/. Specifically, three values of p are introduced, leading to the composite metric $/L_{p1} + L_{p2}/_{p3}$:

$$\varrho/\underline{x},\underline{y}/ = \left[\left(\sum_{i=1}^{n_1} \left| x_i - y_i \right|^{p_1}\right)^{p_3/p_1} + \left(\sum_{i=n_1+1}^{n_1+n_2} \left| x_i - y_i \right|^{p_2}\right)^{p_3/p_2}\right]^{1/p_3} \qquad /19/$$

where $\underline{x}$ and $\underline{y}$ are two vectors, p_1 and p_2 are composite programming parameters for, respectively, the set of n_1 statistical criteria and the set of n_2 observation effort criteria, and p_3 is the parameter for trading off statistical criteria and observation effort. As shown in Theorem 1 using functional analysis /Dunford and Schwarts[13]/ that Equation /19/ defines a metric. The rationale for this seemingly complex formulation is as follows. In general, there are three trade-offs to be made.

1. A trade-off among statistical criteria, which is an analytical operation without directly measurable economic consequence.
2. A trade-off among observation effort criteria, with direct economic consequences.
3. A trade-off between statistical criteria and observation effort criteria, that is, between economic and non-

economic attributes.

It is thus conceivable that the three values of p_1, p_2, p_3 are chosen to be different because of the different nature of the trade-offs. With only one criterion for observation effort, T, one has $n = n_1$, $n_2 = 1$; also, p_2 cancels out in Equation / /, which becomes:

$$CF/\sigma,T/ = \left\{ \alpha_1^{p_3} \left[\sum_{k=1}^{n} \beta_k^{p_1} \left(\frac{\sigma_k - \sigma^*_k}{\max \sigma_k - \sigma^*_k} \right)^{p_1} \right]^{p_3/p_1} + \alpha_2^{p_3} \left[\frac{T^* - T}{T^* - \min T} \right]^{p_3} \right\}^{1/p_3} \quad /20/$$

where $/\alpha_1, \alpha_2/$ are the weights assigned to the trade-off between the two sets of objectives and the β_k are the weights assigned within the statistical criteria, while σ^*_k and T^* are the ideal values.

CF is divided into two parts:

$$CF/\sigma,T/ = \left(\alpha_1^{p_3}\ CF_1^{p_3} + \alpha_2^{p_3}\ CF_2^{p_3} \right)^{1/p_3} \quad /21/$$

where CF_1 is the statistical goal function

$$CF_1 = \left[\sum_{k=1}^{n} \beta_k^{p_1} \left(\frac{\sigma_k - \sigma^*_k}{\max \sigma_k - \sigma^*_k} \right)^{p_1} \right]^{1/p_1}$$

CF_2 is the goal function for observation effort :

$$CF_2 = \frac{T^* - T}{T^* - \min T} .$$

Freimer and Yu[14], and Yu and Leitmann[15] have stated that "parameter p plays the role of the balancing factor between the group utility and the maximum of the individual regrets". With this concept in mind, a distinction is made between estimation accuracy criteria and observation effort criterion T. Since standard deviations σ_k are not physical quantities, an absolute value criteria, which places equal weight on small and large deviations /group utility/ is adequate /and simple/; thus, $p_1 = 1$. The trade-off between accuracy and effort is taken from a least squares viewpoint to provide a good balance between the two types of measures; thus, $p_3 = 2$. Such an approach parallels the multiobjective optimization of river basin plans in Duckstein and Opricovic[16]; for p = 1 the solution emphasized non-economic criteria /especially water quality/ and for p = 2 it leant on economic criteria.

Gradient optimization /Wilde and Beightler[17]/ is used to minimize the statistical goal function CF_1 for a series of fixed values of T.

This minimization results in the trade-off function and network parameters written, respectively as:

$$F/T/ = \min_{a,\Theta} CF_1 /T, a, \Theta//; \Theta^*/T/; a^*/T/.$$

A trade-off between statistical accuracy and observation resource availability is then calculated by minimizing the composite goal function CF over T:

$$\min_T CF = \min \left\{ \alpha_1^2 \left[F/T/-\min F \right]^2 + \alpha_2^2 \, CF_2^2 \right\}^{1/2} \qquad /22/$$

As an example, a two-layer aquifer system of 950 ha typical of Western Hungary, composed of an upper silt layer and lower coarse sand, is considered. Two parameters, thickness and porosity, are to be identified.

The MCDM analysis was performed for a rectangular network corresponding to combinations of areas included within the limits: $2{,}500 \le T \le 5{,}000 \text{ m}^2$.

Table 5

Set of weights for statistical criteria

Parameter		Set of weights				
		I	II	III	IV	V
Silt	thickness	1	1	1	2	2
	porosity	1	2	1	2	3
Coarse Sand	thickness	1	1	2	1	1
	porosity	1	2	2	1	1

Table 6

Results of composite programming

CP Solutions	1	2	3	4	T, m^2
			estm. S.D.		
Weights: I,II,IV,V.					
90m x 40m, $\Theta = 45^0$	0,57	0.068	1,2	0.091	3600
Weights III.					
40m x 90m, $\Theta = 45^0$	0,57	0.079	1,2	0.083	3600

1. Upper-layer: thickness, m
2. Upper-layer: porosity
3. Lower-layer: thickness, m
4. Lower-layer: porosity

Five sets of different weights β were used as indicated in Table 5. The standard deviation of point estimation in the middle of the grid elements was selected as a statistical criterion. The optima of the five single objective functions /Equation /18/ / were taken as ideal values σ^*_k and T^*. Results of composite programming for the five sets of weights β and weights $\alpha_1 = \alpha_2 = 1$ are given in Table 6. The solution is insensitive to changes in β for weight sets I, II, IV and V. When the lower layer receives relatively high preference /set of weights III/, the former solution will rotate by an angle of 90° and results in different standard deviations for porosity estimates.

CONCLUSIONS

The following concluding points can be drawn:

1. It has been demonstrated that composite programming is a proper extension of compromise programming.
2. Composite programming goal functions can be transformed into L_2-type norm functions, thus facilitating the solution procedure.
3. The methodology is especially applicable to problems where conflicting objectives, such as economic, environmental and hydrologic objectives are present, and in addition, to each objective there corresponds one or more criteria.
4. Composite programming appears to be an appropriate technique to model the management of agricultural watersheds.
5. The numerical solution seems to be quite robust in the case when fairly balanced preferences are given to the three objectives of watershed management.
6. Geostatistics and composite programming have been combined in order to find an observation network trading off

estimation accuracy and observation effort.

7. Several parameters exhibiting geostatistical characteristics /spatially correlated and anisotropic/ can be accounted for.
8. The composite programming method applies a compromise objective for the statistical criteria, and another compromise objective for the observation efforts criteria, then seeks for a trade-off between the two compromise objectives.
9. The approach is relatively simple and the algorithm may be implemented on minicomputer.

REFERENCES

1. Zeleny, M., Compromise Programming in Multiple Criteria Decision Making, M.K. Starr and M. Zeleny eds., Univ. of South Carolina Press, 101, 1973.
2. Gershon, M.E., Model Choice in Multiobjective Decision Making in Water and Mineral Resource Systems, Technical Reports, The University of Arizona, Tucson, Arizona, 85721, 155, 1981.
3. Bogárdi, I. and A. Bárdossy, Network design for the spatial estimation of environmental variables, Applied Math. and Comput, 12, 339, 1983.

4. Miller, W.L. and D.M. Byers, Development and display of multiple objective projects impacts, Water Resour. Res., 9 /4/, 11, 1973.

5. Das, P. and Y.Y. Haimes, Multiobjective optimization in water quality and land management, Water Resour. Res., 15 /6/, 1313, 1979.

6. Bogárdi, I., David, L. and L. Duckstein, Trade-off between cost and effectiveness of control of nutrient loading into a waterbody, Research Report, RR-83-19, IIASA, Laxenburg, Austria, July 1983.

7. Williams, J.R., Sediment yield prediction with universal equation using runoff energy factor, Agr.Res.Serv., ARS-S-40, USDA, Washington D.C., 244, 1975.

8. Bárdossy, A., Bogárdi, I. and L. Duckstein, Accuracy of sediment yield calculation, Working paper, Tiszadata, Budapest, 1984.

9. Soil Conservation Service, National Engineering Handbook, Section 4, Hydrology, USDA, Washington D.C., 1971.

10. Bogárdi, I. and A. Bárdossy, A concentration model of P stemming from agricultural watersheds, Research Report /in Hungarian/, Tiszadata, Budapest, 1984.

11. Volpi, G., G. Gambolati, L. Carbognin, P. Gatto and G. Mozzi, Groundwater contour mapping in Venice by

stochastic interpolators, Water Resour. Res., 15 /2/, 291, 1979.

12. Matheron, G., Les variables régionalisées et leur estimation, 306, Masson, Paris, 1967.

13. Dunford, N. and J.T. Schwarts, Linear Operators Part I., Interscience Publisher, New York, 358, 1976.

14. Freimer, M. and P.L. Yu, Some new results on compromise solutions for group decision problems, Management Science, 22 /6/, 688, February 1976.

15. Yu, P.L. and G. Leitmann, Compromise Solutions, Domination Structures, and Szlukyadze's Solution, in Multicriteria Decision Making and Differential Games, edited by G. Leitmann, Plenum Press, New York, 85, 1976.

16. Duckstein, L. and S. Opricovic, Multiobjective optimization in river basin developments, Water Resour. Res., 16 /1/, 14, 1980.

17. Wilde, D.J. and C.S. Beightler, Foundations of Optimization, Prentice-Hall, Inc., 480, 1967.

18. Bárdossy, A., Mathematics of composite programming, Working paper, Tiszadata, Mikó u. 1. 1012. Budapest, Hungary, 1984.

OPTIMIZING THE DISTRIBUTION OF TRADE BETWEEN PORTS AND TRADING CENTRES

A.A. El-Dash
Faculty of Commerce
University of Helwan

J.B. Hughes
Departments of Mathematics
University College of North Wales

Introduction

A problem facing many emerging countries is how to schedule goods to ports and trading centres so as to reduce congestion in their transportation networks.

In this paper we consider a stochastic goal programming model of the situation (confining our attention to a relatively simple transportation model in order to concentrate on its stochastic nature) and indicate how geometric programming techniques may be used to tackle the equivalent deterministic nonlinear goal program.

A Model with chance constraints

Consider a region having M trading centres (labelled i), N ports (labelled j) handling T commodities (labelled t). We define the following quantities

A_t = total amount of commodity t at the centres to be transported to the ports, per unit time.

B_t = total amount of commodity t at the ports to be transported to the centres, per unit time.

L_{jt} = handling capacity of port j for commodity t , per unit time.

X_{ijt} = amount of commodity t allocated by centre i to port j , per unit time.

Y_{ijt} = amount of commodity t allocated by port j to centre i , per unit time.

C_{ijt} = cost of transporting unit amount of commodity t between centre i and port j .

C_t = the cash limit for transporting commodity t , per unit time.

d_{ijt} = handling capacity of the transport link between centre i and port j for commodity t , per unit time.

We consider the case when the quantities A_t , B_t and C_{ijt} are non negative random variables and the constraints in which they appear to be chance constraints. This requires the decision maker to ascribe tolerance measures

α_t , β_t , γ_t reflecting 'acceptable' levels for the violation of the chance constraints involving A_t , B_t and C_{ijt} respectively.

For our model we consider the following set of constraints, which is adequate for present purposes,:

$$Pr\ (\sum_{i,j} X_{ijt} \geq A_t) = \alpha_t \tag{2.1}$$

$$Pr\ (\sum_{i,j} Y_{ijt} \geq B_t) = \beta_t \tag{2.2}$$

$$Pr\ (\sum_{i,j} C_{ijt}\ (X_{ijt} + Y_{ijt}) \leq C_t) = \gamma_t \tag{2.3}$$

$$\sum_{i} (X_{ijt} + Y_{ijt}) \leq L_{jt} \tag{2.4}$$

$$\sum_{ij} (X_{ijt} + Y_{ijt}) \leq d_t \tag{2.5}$$

Equivalent deterministic constraints

We replace constraints (2.1) and (2.2) by the following deterministic constraints:

$$\sum_{ij} X_{ijt} = F_{A_t}^{-1} (\alpha_t) \tag{3.1}$$

$$\sum_{ij} Y_{ijt} = F_{B_t}^{-1} (\beta_t) \tag{3.2}$$

where F_{A_t} and F_{B_t} are the c.d.f.'s for A_t and B_t respectively.

To obtain a deterministic constraint equivalent to (2.3) we have to make assumptions about the non-negative random variables C_{ijt} . We consider that each can be approximated by a chi square distributed random variable with an even number of degrees of freedom $2g_{ijt}$. This allows us to use a result due to Box[1,2].

Box's Theorem: If $\chi^2(2g_k)$ is a chi square distributed variable with $2g_k$ degrees of freedom and λ_k is a constant, the exact distribition of

$$y = \sum_{k=1}^{n} \lambda_k \chi^2 (2g_k) ,$$

where g_k is an integer, is given by

$$\Pr (y > y_0) = \sum_{k=1}^{n} \sum_{s=1}^{g_k} \eta_{ks} \Pr \left[\chi^2(2s) > y_0/\lambda_k\right]$$

where each η_{ks} is a constant involving only the λ's. Using Box's theorem, (2.3) is equivalent to

$$\sum_{i,j} \sum_{s=1}^{g_{ijt}} \eta_{ijst} F_{ijt} \left[\frac{C_t}{x_{ijt} + y_{ijt}}\right] = 1 - \gamma_t \qquad (3.3)$$

where F_{ijt} is the c.d.f. of a chi-square distribution with $2g_{ijt}$ degrees of freedom and the coefficients η_{ijst} are rational functions of $x_{ijt} + y_{ijt}$. Constraint (3.3) is highly nonlinear in the decision variables x_{ijt} and y_{ijt}. The form of the coefficients η_{ijst} is suitable for using geometric programming but F_{ijt} is an incomplete gamma function. The standard geometric programming technique for terms with exponential functions is to write

$$e^x \approx \left[1 + \frac{x}{\phi}\right]^{\phi} , \text{ where } \phi \to \infty , \qquad (3.4)$$

and we write $\Phi_{ijt}(\ ,\phi)$ for $F_{ijt}(\ )$ when this approximation is used.

The constraints of the model, with non negative deviational variables included in each, are then:

$$\sum_{ij} X_{ijt} + x_t^- - x_t^+ = F_{A_t}^{-1}(\alpha_t) \tag{3.5}$$

$$\sum_{ij} Y_{ijt} + y_t^- - y_t^+ = F_{B_t}^{-1}(\beta_t) \tag{3.6}$$

$$\sum_{ij} \sum_{s=1}^{g_{ijt}} \eta_{ijst} \Phi_{ijt} [C_t, X_{ijt} + Y_{ijt}, \phi] + C_t^- - C_t^+ = 1 - \gamma_t \tag{3.7}$$

$$\sum_{i} [X_{ijt} + Y_{ijt}] + \ell_{jt}^- - \ell_{jt}^+ = L_{jt} \tag{3.8}$$

$$\sum_{ij} [X_{ijt} + Y_{ijt}] + d_t^- - d_t^+ = d_t . \tag{3.9}$$

A nonlinear goal programming model

Given a preemptive priority structure for the optimization, say:

First: to minimize $\sum_t \left[x_t^- + y_t^-\right]$, the amount of goods not allocated for transportation.

Second: to minimize $\left[\sum_{jt} \ell_{jt}^+ + \sum_t d_t^+\right]$ the total amount of goods allocated that cannot be handled at the ports or transported between ports and centres.

Third: given tolerance levels γ_t, for the probabilities that transport goals are exceeded, to minimize $\sum_t C_t^-$ the amount by which the sum of the actual probabilities exceeds the sum of the tolerance levels

the model yields a non linear goal program.

For the above priority structure, it is:

Find X_{ijt} , $Y_{ijt} \geq 0$ so as to

$$\text{lexico min} \quad a = \left\{ \sum_t \left[x_t^- + y_t^- \right] , \left[\sum_{jt} \ell_{jt}^+ + \sum_t d_t^+ \right] , \sum_t C_t^- \right\} \quad (4.0)$$

subject to

$$\sum_{ij} X_{ijt} + x_t^- \geq F_{A_t}^{-1} \left[\alpha_t\right] \quad (4.1)$$

$$\sum_{ij} Y_{ijt} + y_t^- \geq F_{B_t}^{-1} \left[\beta_t\right] \quad (4.2)$$

$$\sum_{ij} \sum_{s=1}^{g_{ijt}} \eta_{ijst} \, \Phi_{ijt} \left[C_t, X_{ijt} + Y_{ijt}, \phi\right] + C_t^- \geq 1 - \gamma_t \quad (4.3)$$

$$\sum_i \left[X_{ijt} + Y_{ijt}\right] - \ell_{jt}^+ \leq L_{jt} \quad (4.4)$$

$$\sum_{ij} \left[X_{ijt} + Y_{ijt}\right] - d_t^+ \leq d_t \, . \quad (4.5)$$

(Minimizing (4.0) ensures that at optimality constraints (4.1) - (4.5) will be equalities and equivalent to (3.5) - (3.9) since $x_t^+ = y_t^+ = C_t^+ = \ell_{jt}^- = d_t^- = 0$.)

The problem is solved using the sequential goal programming algorithm due to Dauer and Krueger[3] in which, for each priority level, the optimization is carried out using the double condensation algorithm for geometric programming due to Avriel, Dembo and Passy[4] which is outlined in the Appendix.

It should be noted that we assume that a solution obtained using approximation (3.4) converges to the true solution as $\phi \to \infty$. In practice, the problem is solved on a computer using a suitably large value for ϕ .

Appendix

<u>The arithmetic-geometric mean inequality</u> For $u_t\ ,\ \delta_t > 0\ ,\ \sum_{t=1}^{m} \delta_t = 1$

$$\sum_{t=1}^{m} u_t \geq \prod_{t=1}^{m} \left[\frac{u_t}{\delta_t}\right]^{\delta_t} \tag{A1}$$

<u>Posynomials and condensed posynomials</u> A posynomial $P(x)$ of $x = (x_1, x_2, \ldots x_n)$ is a function of the form

$$P(x) = \sum_{t=1}^{m} p_t(x) \text{ where } p_t(x) = C_t x_1^{a_{1t}} \ldots x_n^{a_{nt}},\ C_t > 0.$$

Using (A1) with $u_t = p_t(x)\ ,\ \delta_t = \dfrac{p_t(\bar{x})}{P(\bar{x})}$

$$P(x) = \sum_{t} p_t(x) \geq \prod_{t}^{m} \left[\frac{p_t(x)}{\delta_t}\right]^{\delta_t} = P(\bar{x}) \prod_{r=1}^{n} \left[\frac{x_r}{\bar{x}_r}\right]^{h_r(\bar{x})} = P(x,\bar{x})$$

$P(x,\bar{x})$ is the condensed posynomial of $P(x)$ at the point $\bar{x}$. It is a monomial in x and is such that

$$P(x) \geq P(x,\bar{x}) \quad \text{with equality when} \quad x = \bar{x} \,. \tag{A2}$$

Signomials A signomial $G(x)$ in x is any function of x which can be expressed as the difference of two posynomials in x .

The double condensation algorithm In what follows, G denotes a signomial, P and Q denote posynomials and $\bar{P},\bar{Q}$ denote the condensed posynomials corresponding to P,Q when the condensation has been carried out about a point $\bar{x}$ which is feasible for programs 1 and 2 below.

PROGRAM 1

$$\min \; G_0$$

$$\text{subject to} \quad G_i \leq \sigma_i \,, \quad \sigma_i \in R \,, \; i = 1,2,\ldots m$$

$$x > 0$$

PROGRAM 2

$$\min \; x_0$$

$$\text{subject to} \quad \frac{P_i'}{Q_i'} \leq 1 \qquad i = 0,1,2,\ldots m$$

$$x_0 \,, \; x > 0$$

LEMMA Programs 1 and 2 are equivalent.

PROGRAM 3

$$\min \; x_0$$

$$\text{subject to} \quad \left[\bar{Q}_i'\right]^{-1} P_i' \leq 1 \qquad i = 0,1,2,\ldots m$$

$$x_0 \,, \; x > 0$$

PROGRAM 4

$$\min \; x_0$$

$$\text{subject to} \quad \overline{\left[\bar{Q}_i'\right]^{-1} P_i'} \leq 1 \qquad i = 0,1,2,\ldots m$$

$$x_0 \ , \ x > 0$$

PROGRAM 5 $\qquad \min \ z_0$

$$\text{subject to} \quad \sum_{j=0}^{n} a_{ij} \, z_j \le b_i \qquad i = 0,1,\ldots m$$

$$0 \le z_i \le \bar{z}_i$$

Program 5 is equivalent to Program 4, amd can be derived from it using

$$z_j = \ln\left[\frac{X_j}{(x_j)_{LB}}\right] \quad , \quad \bar{z}_j = \ln\left[\frac{(x_j)_{UB}}{(x_j)_{LB}}\right]$$

where $0 < (x_j)_{LB} \le x_j \le (x_j)_{UB}$ [$(x_j)_{LB}$, $(x_j)_{UB}$ have to be specified.] Program 5 turns out to be linear program because the constraints in Program 4 are monomials.

Dembo[5] has proved

(i) the solution vector for Program 5 is infeasible for Program 2

(ii) $\min z_0$ (Program 5) $\le \min x_0$ (Program 2).

These observations are the basis of a cutting-plane algorithm for solving Program 2 in which Program 5 is solved repeatedly with an additional constraint (cut) included at each iteration which reduces the feasible region of Program 5 without affecting that for Program 2. These solutions to Program 5 converge to a feasible point for Program 2 which will be (at least) a local minimum if the Kuhn-Tucker conditions are satisfied at that point.

The method depends on finding an initial feasible point x for program 2 about which the condensations are performed. This may not be a trivial problem. Avriel, Dembo and Passy[4] give also another algorithm (based on their double condensation algorithm) which, if certain conditions are satisfied, is guaranteed to yield a feasible point for Program 2.

This second algorithm may also be used to attempt to improve on a solution of Program 2 obtained using the first algorithm and which is a local but not a global minimum.

References

1. Box, G.E.P., Some theorems on quadratic forms applied in the study of analysis of variance problems, 1. Effect of inequality of variance in the one-way classification, *Ann. Math. Stat.*, 25 No2, 290, 1954.

2. El-Dash, A.A., Chance constrained and non-linear goal programming, *Ph.D. Thesis*, University of Wales, 1984.

3. Dauer, J.P. and Krueger, R.J., An iterative approach to goal programming, *Operational Research Quarterly*, 28 No3, 671, 1977.

4. Avriel, M., Dembo, R. and Passy, U., Solution of generalized geometric programs, *International Journal for Numerical Methods*, 9, 149, 1975.

5. Dembo, R.S., Solution of complementary geometric programming problems, *M.Sc. Thesis*, Technion - Israel Institute of Technology, Haifa, Israel, 1972.

Bibliography

Avriel, M., (editor), Advances *in geometric programming*, Plenum Press, New York, 1980.

Beightler, C.S. and Phillips, D.T., *Applied geometric programming*, John Wiley and Son Inc., 1976.

AIDS FOR MULTI OBJECTIVE DECISION MAKING ON DEVELOPMENT OF TRANSPORTATION NETWORK UNDER UNCERTAINTY

Miroslaw Bereziński, Lech Krus
Systems Research Institute
Polish Academy of Sciences

INTRODUCTION

The decision making problem considered in the paper deals with development planning of transportation system (TS). The planning process has two-level hierarchical structure. On the higher level (level of national economy planning performed at planning commitee) only decision concerning expected and desired macroproperties of the system are undertaken. Inputs and outputs of the system are considered in aggregated way. On the lower level (performed at communication department) one undertakes decisions relating to the manner of utilisation of the inputs as well as these concerning the organization of traffic flows in the network, planning of work organization, network management etc. The lower decisions are supervised and coordinated by the higher level. In the paper only the higher level planning problem is considered and an approach is proposed aiding the problem. Let the planning interval be T, the required final state vector of TS - x^* the planned inputs to TS - u(t), the TS final state - x(T). The final state is treated as objective vector. The required state results from the demand of national economy on the transportation services. The planning problem consists in looking for such control variables u that the final state x(T) at the planning interval T will be possibly close to the required vector x^*.

Process of planning of the transportation system development is realized under uncertainty due to the incompletness and incertitude of statistical data, uncertainty introduced by the decision makers, uncertainty and incompletness of the knowledge about the future state of the national economy and future technical features of the system etc. The sources of uncertainty in the problem can be classified as: uncertain inputs and disturbances of TS system, internal noises, uncertain required future TS state.

STOCHASTIC MACRO MODEL OF TRANSPORTATION SYSTEM

In the presented model see fig.1 two input streams are

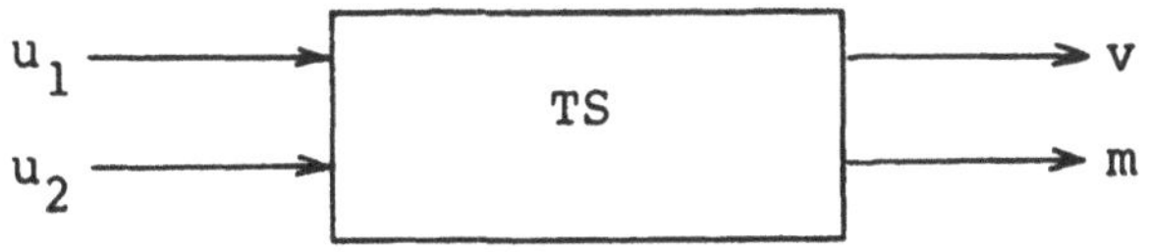

Fig.1. Macro model of transportation system.

considered. The first one named an organization stream refers the process of inflow into TS of such factors as manpower, technical ideas, new concepts of transportation means, new organization forms etc. It is measured by u_1 the entropy which characterizes the level of disorder of the transport processes. The second one u_2 is called a stream of energomatery and includes production means such as energy, new transportation means, investments for infrastructure (reailway, its equipment, and buildings etc.). Both the stream of energomatery and that of organization are not strictly determined. Therefore, it seems to be reasonable to assume that thay can be reflected in mathematical form as suitable stochastic processes. We assume that these processes are identificable i.e. we are able to estimate their probabilistic properties by the means of appropriate statistical methods.

The output quantities of TS are the transportation network capacity m and average speed v of realization of transportation processes. The capacity and the average speed characterize the state of TS. Initial state $m(0)$ and $v(0)$

is given and determined. The model can be described (see Bereziński and Krajewski [1]) by the following state equations in descrete form:

$$\Delta m(t)=a_{11}m(t)+a_{12}v(t)+b_{11}u_1(t)+b_{12}u_2(t) \tag{1}$$

$$\Delta v(t)=a_{21}m(t)+a_{22}v(t)+b_{21}u_1(t)+b_{22}u_2(t) \tag{2}$$

The matrices $A=(a_{ij})$ and $B=(b_{ij})$ are assumed to be determined, but the input streams we treat as random processes.

Let us denote the vectors $x(t)=(m(t),v(t)), u(t)=(u_1(t), u_2(t))$. The expected values of the output $\overline{x}=(Em(t),Ev(t))$ can be derived:

$$\overline{x}(t)=e^{At}\overline{x}(0)+\sum_{\tau=0}^{t} e^{A(t-\tau)}B\,\overline{u}(\tau). \tag{3}$$

The solution of the model for random variables $x(t)$ takes the form:

$$x(t)=\overline{B}\cdot\overline{u}+\overline{A}+n \tag{4}$$

where

$\overline{u}=\{\overline{u}_1(0),\overline{u}_1(1),\ldots,\overline{u}_1(t-1),\overline{u}_2(0),\overline{u}_2(1),\ldots,\overline{u}_2(t-1)\}$,

$\overline{B}$ matrix and $\overline{A}$ vector are obtained from (3),

n is a random variable with the expected value $E(n)=0$ and the covariance matrix

$$R_x(t)=\sum_{\tau_2=0}^{t}\sum_{\tau_1=0}^{t} e^{A(t-\tau_1)}BR_u(\tau_1,\tau_2)B^T e^{A^T(t-\tau_2)}, \tag{5}$$

where R_u is the correlation matrix of the input variables $u(t)$, $(\cdot)^T$ denotes transposition.
Assuming given form of probability distribution of random variables representing input process, the probability distribution of the variable representing output process can be generally calculated.

MULTICRITERIAL OPTIMIZATION PROBLEM

For given planning interval T we have from the model solution

$$x(T) = \overline{B}\,\overline{u} + \overline{A} + n \,. \tag{6}$$

We assume linear constraints on inputs - control variables $\overline{u}$

$$\overline{u} \in V = \{\overline{u} : C\,\overline{u} < D,\ \overline{u} > 0\} \tag{7}$$

where the matrix C and the vactor D are given.
The goal x^* to be reached by $x(T)$ at the planning interval T is defined by the expected value $\overline{x}^* = E(x^*)$ and a probability distribution $x(T), n, x^*$ are treated as random variables. The gaussian distributions of random variables representing, the input process and the goal to be reached are assumed. The processes are assumed to be statistically independent.

The problem consists in looking for such inputs $\overline{u}$ that the output $x(T)$ of the TS will reach the goal x^* at the planning interval T with the maximum probability.

Considering two sources of uncertainty, the first of the input stream and the second of the goal to be reached, we have in generally that the output $x(T) \neq \bar{x}(T) = \bar{B}\bar{u} + \bar{A}$ and that the goal $x^* \neq \bar{x}^*$. In these conditions we choose a region $W(f_1, f_2, \alpha)$ depending on: - the probability densities f_1 attached to the output $x(T)$ and f_2 attached to the goal x^* - and the parameter α proposed by planner. The region W^* is choosen in the objectives space R^2 such that $\bar{x}^* \quad W^*$. We look for such input $\bar{u}$ that $x(T)$ has the highest probability of falling in the region W^*:

$$\max_{\bar{u} \in U} P\{x(T) \in W^*(f_1, f_2, \alpha)\}. \tag{8}$$

To describe the region W^* let us consider first the densities f_1 and f_2. The qaussian denisity of the output $x(T)$ has the form:

$$f_1 = (2\Pi)^{-1} |R_x|^{-1/2} e^{-Q_1/2} \tag{9}$$

where the quadratic form $Q_1 = (x(T) - Bu - A)^T R_x^{-1} (x(T) - Bu - A)$, R_x is the covariance matrix (5) for $t=T$.
The gaussian density of the goal x^* has the form

$$f_2 = (2\Pi)^{-1} |R_2|^{-1/2} e^{-Q_2/2} \tag{10}$$

with $Q_2=(x^*-\bar{x}^*)^T R_2^{-1} (x-\bar{x}^*)$, $\bar{x}^*$ and R_2 are estimated by statistical treatment of the information given by experts. We introduce new random variable w with the expected value $\bar{w}=\bar{x}^*$ and with the density f being linear combination of f_1 and f_2:

$$f(w) = \alpha \cdot f_1(n) + (1-\alpha) f_2(x^*), \quad \alpha \in [0,1]. \tag{11}$$

The density f is gaussian and has the form

$$f(w) = (2\pi)^{-1} |R_w|^{-1/2} e^{-Q_w/2} \tag{12}$$

where the covariance matrix R_w, is a function of R_x, R_2, α and the quadratic form Q_w is as follows:

$$Q_w = (w - \bar{x}^*)^T R_w^{-1} (w - \bar{x}^*). \tag{13}$$

The parameter α defines the mass proportion of probabilities described by f_1 and f_2, taken into account in $f(w)$. The region W^* is constructed in such a way that for some constant c it is defined by an elipse described by the form Q_w:

$$W^* = \{w=(w_1,w_2) : Q_w=(w-\bar{x}^*)^T R_w^{-1} (w-\bar{x}^*) < c\} \tag{14}$$

For nonsingular R_w the form Q_w has χ^2 distribution. The region W^* can be interpreted as a confidence region for $x(T)$ at some level β. Let for some u the expected value $E(\overline{Bu} + \bar{A} + n)=\bar{x}^*$, then the probability $P(x(T) \in W^*) = \beta$.

It can be shown (similary as by Contini.[2]) that the problem (8) is equivalent with the following quadratic programming model:

$$\min_{u \in U} (\overline{x}^* - \overline{Bu} - \overline{A})' R_W^{-1} (\overline{x}^* - \overline{Bu} - \overline{A}). \tag{15}$$

It can be solved by known programming codes.

The considered approach is presented in case of two dimentional space objectives. It is however valid for more, say m, dimentional spaces. All the formulae (6-15) are correct in such case. The region W^* would have the form of m dimentional elipsoide.

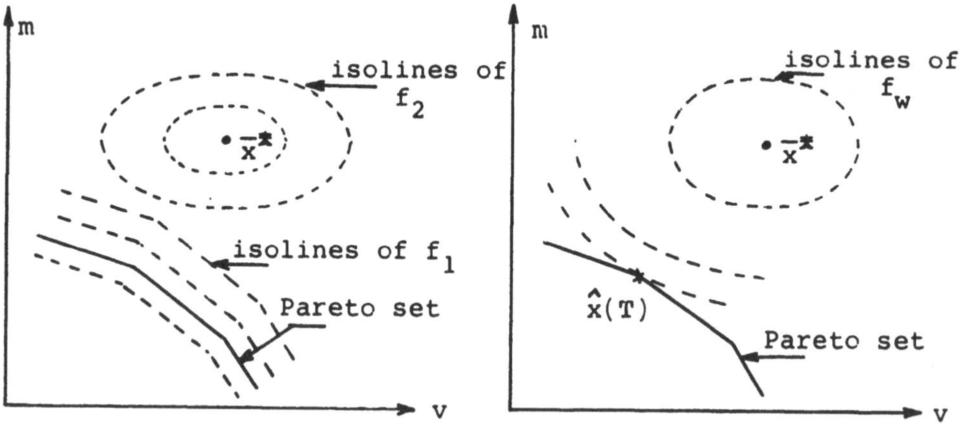

Fig.2. Illustration of the probability distributions f_1, f_2, f_W in the space of objectives.

An illustration is given by the fig.2 in the space of objectives. On the left side the Pareto set for the model $\overline{x}(T)=$

$=\overline{Bu}+\overline{A}$ subject to the constraints $C\overline{u} < D$ is presented together with isolines (dropped) of the probability distribution f_1. The desired goal is shown in form of expected value $\overline{x}^*$ and isolines (dropped) of the distribution f_2. On the right side the uncertainty of the model is shifted to the expected value $\overline{x}^*$ and combined together with the uncertainty of the goal. The goal $\overline{x}^*$ is given together with elipses - isolines (dropped) of the distribution f_w. The problem is boiled down to looking for minimal distance between the expected value of goal and the Pareto set in the sense of elipse norm.

EVALUATION OF GOALS BY EXPERTS METHOD

The prognostic graph (PG) method which has been elaborated by Gluszkov [3] is proposed for evaluation of the forecased goals. The method joins ideas of two american methods: DELPHI and PERT. The PG-method utilises both experts' procedures and mathematical algorithms which are used in the sequential process of prognostic graph constructing.

The procedure consists of the three following steps:
STEP 1: On the basis of national economy forecast and its demand for transportation services an expert group formulates

a set of candidates for final goals $s_1, s_2, \ldots, s_m$, which should be reached by transportation system in the period T. Taking into account a relative importance of the candidates the experts attach to each candidate s_i appropriate in their opinion weights γ_i where $i=1,2,\ldots,m$. This is done by a permanently acting group of experts which is in contact with the department of comunication.

STEP 2: An other group of experts proposes an initial list of intermediate events and constructs a prognostic graph with the events as nodes and with interactions between them - as vertices. To do it each expert must indicate a set of intermediate events which in his opinion should absolutely occure before appearing of the particular candidate. The procedure is iterative. At each iteration new intermediate events are added and new versions of prognostic graph are constructed. The procedure terminates when the experts can not formulate any new intermediate event. As a result one obtains a list of all intermediate events and various versions of the prognostic graph.

STEP 3: The obtained versions of prognostic graph are assumed to be realizations of stochastic graph. They are processed by the use of appropriate statistical techniques. As a result one obtaines average values of forecased goals as well as estimates of unknown parameters of probability distribution.

CONCLUSIONS

In the paper the approach is proposed aiding multicriterial planning of development of transportation system on the level of national economy. This macro approach should be used in practice together with micro approach (not considered here) (on the level of communication department).

The approach consists in application of linear stochastic model describing development of the system and in utilisation of the experts method for evaluation of the goals to be reached. Unicertainties about the input stream of the transportation system has been taken into account as well as uncertainty about the evaluated goals. For normal probability distributions of random processes describing inputs and goals, the solution is obtained in form of quadratic programming problem. The planner has possibility to define the mutual ratio of probability masses connected with randomness of the model output and uncreainty of the goal evaluation taken into accunt in the model.

REFERENCES

1. BEREZINSKI M., KRAJEWSKI W.: MACROSCOPE MODELLING OF RAILWAY NETWORK: THEORETICAL APPROACH AND EXAMPLES OF MODEL VERIFICATION ON THE NATIONAL RAILWAY NETWORK. RESEARCH PA-

PER, TECHNICAL UNIVERSITY OF WARSAW, TRANSPORT INSTITUTE, WARSAW 1983 (in polish).

2. CONTINI B.: A STOCHASTIC APPROACH TO GOAL PROGRAMMING. OPERATIONS RESEARCH, 16 (3), 576-586, 1983

3. GLUSZKOV M.: METHODOLOGY OF SCIENCE AND TECHNICS DEVELOPMENT FORECASTING, KIEV 1971 (in russian).

SCALARIZATION AND THE INTERFACE WITH DECISION MAKERS IN INTERACTIVE MULTI OBJECTIVE LINEAR PROGRAMMING

Matthijs Kok
Department of Mathematics and Informatics
Delft University of Technology

ABSTRACT

Interactive multi-objective linear programming methods are based on three different types of scalarizing functions. In this note we discuss the amount of information to be given by decision makers for each of these types. We show that in many methods the ability of a decision maker to oversee a large number of stimuli is overestimated.

1. INTRODUCTION

The most important assumption for Multi-Objective Linear Programming (MOLP) methods is that a decision maker or a group of decision makers exists who is able to give <u>preference information</u> in order to obtain a final compromise solution. This solution is not, as in classical single-objective optimization, determined by necessary and sufficient conditions formulated in mathematical terms. It is determined by the subjective preferences of the decision makers. It is fashionable, today, to talk about interactive decision making (whereby the speaker refers to interactions between the model and the decision makers), as if a new concept is presented. Instead, modelling has always been interactive, thus providing a trial and error method to obtain more insight in a given problem. Real-time decision making (with real-time interaction between the model and the decision makers) is not new either, but new hardware developments (mini computers, personal computers, computer networks) steadily increase the possibilities for previously unfeasible ways of decision making. In this note, however, interactive means "progressive articulation of preferences". The crucial, non-mathematical, question now is how to obtain the preference information and how to constitute the man-machine procedure. All proposed methods differ widely in this respect. We will review a broad class of methods by investigating the underlying mathematical formulation. It turns out that the interactive methods are based on three concepts: the weighting method, the constraint method and the reference-point method (section 2). The amount of information to be assessed by the decision maker(s) in each of these interactive methods is investigated in section 3. In section 4 we will give some final remarks.

2. SCALARIZING FUNCTIONS

The MOLP problem can be formulated as:

$$\begin{aligned} &\max z \\ &\text{subject to} \quad z - Cx = 0 \\ &\qquad\qquad\qquad Ax = b \qquad\qquad (1) \\ &\qquad\qquad\qquad\ \ x \geq 0 \end{aligned}$$

with z the p-vector of goal variables, x the n-vector of decision variables, C the p x n-matrix and A the m x n-matrix of coefficients, b the m-vector of right-hand sides. The three different types of scalarization are:

a. The weighting method. We consider a linear combination of the goal variables with a weighting vector $\lambda \in \mathbb{R}^p$, such that $\lambda_i \geq 0$, $i = 1,2, \ldots, p$, and $\sum_{i=1}^{p} \lambda_i = 1$.

The MOLP problem can now be formulated as the single-objective LP problem:

$$\begin{aligned} \max \quad & \lambda^T z \\ \text{s.t.} \quad & z - Cx = 0 \\ & Ax = b \\ & x \geq 0 \end{aligned} \qquad (2)$$

The Zionts-Wallenius[1] method is based on this formulation.

b. The constraint method. We optimize one of the goal variables subject to the additional constraints that the remaining goal variables are not below given lowerbounds. The resulting single-objective problem reads:

$$\begin{aligned} \max \quad & z_k \\ \text{s.t.} \quad & Ax = b \\ & z_i \geq \ell_i, \quad i = 1,2, \ldots, p, \quad i \neq k \\ & z - Cx = 0 \\ & x \geq 0 \end{aligned} \qquad (3)$$

The Interactive Multiple Goal Programming method (Spronk [2]) and the Interactive Surrogate-Worth Tradeoff method (Chankong and Haimes [3]) are based on this formulation.

c. The reference-point method. We seek a goal vector with smallest distance from a reference vector. The problem reads

$$\min\{ \max_{i=1,..,p} \lambda_i(\tilde{z}_i - z_i)\} - \varepsilon \sum_{i=1}^{p} z_i$$

$$\text{s.t.} \quad Ax = b$$
$$z - Cx = 0 \qquad (4)$$
$$x \geq 0$$

with ε a small positive number and $\tilde{z}$ the ideal (reference) vector. Note that we can use other metrics; e.g. $\Sigma\, \lambda_i\, (\tilde{z}_i - z_i)$ which is equivalent with (2). Interactive methods based on this scalarization (sometimes combined with the constrained method) are: the Reference-Point method (Wierzbicki [4]), the weighted augmented Tchebycheff method (Steuer and Choo [5]), the refined Displaced Ideal method (Zeleny [6]), the STEP-method (Benayoun e.a. [7]) and the Pairwise Comparisons method (Kok and Lootsma [8]). An extensive treatment of the theory behind these scalarizations can be found in Kok [9].

3. INTERFACE WITH DECISION MAKERS

Many psychological studies have shown that human capabilities in processing information are limited (e.g. "In performing complex tasks individuals utilize different heuristics that will keep the information processing demands of the situation within the bounds of their limited capacity", Payne[10]). Moreover, as the amount of information given to the decision maker increases, the percentage of information used decreases (Payne[10]). Especially in medium size or large decision problems this can result in unreliable answers of the decision makers. The number of items a decision-maker has to assess simultaneously and per iteration is listed in table 1 (we will not discuss the number of iterations, because this mainly depends on the capabilities of the decision maker). From this table we can conclude that preference responses in methods 1, 2 and 4 may be unreliable, since they are given using only a part of the available information. Furthermore, methods 6 and 8 are unsatisfactory since these methods include <u>partial</u> tradeoffs. The "stepsize" in method 6 in the direction of an attractive tradeoff is very difficult to specify. So, among the eight methods we have investigated the methods 3,5, 7 (and perhaps 8) are most promising.

4. FINAL REMARKS

Solutions of the weighting method are always basic solutions. Interactive methods that produce basic solutions only are in general too restrictive, thus it is necessary to include a procedure to explore the efficient faces. This means that the constraint method and the reference-point-method have a big advantage with respect to the weighting method. The main problem of interactive MOLP is to develop a methodology such that decision makers have much insight in the decision problem, e.g. by providing information on request.

METHOD	EVALUATION SIMULTANEOUSLY (number of items)	EVALUATION IN ONE ITERATION (number of items)	ASSUMPTIONS
1. Zionts-Wallenius method	12	120	10 efficient tradeoff-vectors
2. Augmented Tchebycheff method	30	30	minimal 5 solutions offered
3. Displaced Ideal method	2	12	no feedback in this method
4. Reference-Point method	12	12	
5. Interactive Multiple Goal Programming Method	7	7	with feedback
6. Surrogate-Worth Tradeoff method	2 + 6	10 + 6	6 items must be assessed for stepsize
7. STEP method	2	12	no feedback in this method
8. Pairwise Comparison method	2	30	

Table 1. Number of evaluations carried out by a decision maker in some interactive methods in a medium size decision problem (p=6, n=500, m=200).

REFERENCES

1. Zionts, S. and J. Wallenius. An interactive multiple objective linear programming method for a class of underlying nonlinear functions, *Management Science*, 1983, vol. 29, no. 5, p. 519-529.
2. Spronk, J., *Interactive Multiple Goal Programming; Applications to financial Planning*, 1981. Martinus Nijhof, Boston.
3. Chankong, V., and Y.Y. Haimes. The interactive Surrogate-Worth Trade-off (ISWT) method for multiobjective decision-making, in S. Zionts (ed.), *Multiple Criteria Problem Solving*, Springer, Berlin, etc., 1978, p. 42-67.
4. Wierzbicki, A.P.. *The use of reference objectives in multi objective optimization. Theoretical implications and practical experiences.* IIASA, WP 79-66, 1979, Laxenburg, Austria.
5 Steuer, R.E., and E.U. Choo, An interactive weighted Tchebycheff procedure for multiple objective programming, *Mathematical Programming*, 26, 1983, p. 326-344.
6. Zeleny, M., *Linear Multi-Objective Programming*, Springer, Berlin, 1974.
7. Benayoun, R., Montgolfier, J. de, Tergny, J., and Larichev, O.I., Linear Programming with multiple objective functions: STEP method (STEM), *Mathematical Programming 1*, 1971, p. 366-375.
8. Kok, M. and F.A. Lootsma, 1984. Pairwise-comparison methods in multiobjective programming, with applications in a long-term energy-planning model. *Report of the Department of Mathematics and Informatics*, 84-19, Delft.
9. Kok, M., The interface with decision makers in interactive multiobjective linear programming methods, *Report of the Department of Mathematics and Informatics*, 1984, Delft.
10. Payne, J.W.. Task complexity and contingent processing in decision making: an information search and protocol analysis, *Organizational Behavior and Human Performance*, 16, 1976, p. 366-387.

LIST OF PARTICIPANTS

Giovanni ANDREATTA, LADSEB-CNR, Corso Stati Uniti 4, 35100 Padova, Italy.

Charalambos BANIOTOPOULOS, Institute for Steel Structures, Aristotle University, Thessaloniki 54636, Greece.

Mihai BERCEA, Institut Polytechnic Iasi, Fac. de Mecanica Catedra "Organe de Masini", Splai Bahlui m. 62, 6600 Iasi, Romania.

Miroslaw BEREZINSKI, System Research Institute, Polish Academy of Sciences, Newelska 6, 01-447 Warszawa, Poland.

Istvan BOGARDI, Tuzadata, Miko u. 1, 1012 Budapest, Hungary.

Ennio CAVAZZUTI, Istituto Matematico, Università di Modena, Via G. Campi 213 B, 41100 Modena, Italy.

Chandan CHOWDHURY, Dept. of Precision Mechanics and Optics, Technical University of Budapest, Egry Jozsef u. 1. III Em., 1521 Budapest, Hungary.

Michel CRISTESCU, CETIM, B.P. 67, 60304 Senlis, France.

Gianni DAL MASO, Istituto di Matematica, Informatica e Sistemistica, Università di Udine, Italy.

Jens FROMM, DFVLR, Linder Hoehe, D-5000 Koeln 90, Western Germany.

Roger HARTLEY, Dept. of Decision Theory, University of Manchester, Manchester M13 9PL, Great Britain.

Mordechai HENIG, Faculty of Management, Tel Aviv University, Tel Aviv 69978, Israel.

Barrie HETHERINGTON, Manchester Business School, Booth Street West, Manchester M15 6PB, Great Britain.

Harro HEYER, Inst. f. Strömungsmechanik u. Elektr. Rechnen im Bauwesen, Universität Hannover, 3000 Hannover 1, Western Germany.

Manfred HUBER, Strada del Vino 46, 39040 Termeno, Bolzano, Italy.

John B. HUGHES, Department of Applied Mathematics, U.C.N.W., Bangor, Gwynedd, LL 57 2UW Great Britain.

Edward M. IBRAHIM, Department of Mathematics, Faculty of Engineering, Ain Shams University, Abassia, Cairo, Egypt.

Johannes JAHN, Fachbereich Mathematik, Schlossgarten Strasse 7, Technische Hochschule Darmstadt, 6100 Darmstadt, Western Germany.

Matthijs KOK, Department of Mathematics and Informatics, Delft University of Technology, Julianalaan 132, 2628 BL Delft, Netherlands.

Lech KRUS, Systems Research Institute, Polish Academy of Sciences, ul. Newelska 6, 01-447 Warsaw, Poland.

Jacek KRUZELECKI, Inst. of Mechanics and Machine Design, Technical University of Cracow, ul. Warszawska 24, 31-155 Cracow, Poland.

Roberto LUCCHETTI, Dipartimento di Matematica, Università di Milano, Via Saldini 50, 20133 Milano, Italy.

Francesco MASON, Dipartimento di Matematica Applicata e Informatica, Università Ca' Foscari, Dorsoduro 3861, Venezia, Italy.

Sandor MOLNAR, Central Inst. for the Development of Mining, Mikoviny s.u. 2-4, H-1037 Budapest, Hungary.

Hirotaka NAKAYAMA, Department of Applied Mathematics, Konan University, 8-9-1 Okamoto, Higashinada, Kobe 658, Japan.

Subhash NARULA, School of Business, V.C.U., 1015 Floyd Avenue, Richmond, Virginia 23284, USA.

Nicoletta PACCHIAROTTI, Istituto Matematico, Università di Modena, Via Campi 213/B, 41100 Modena, Italy.

Adriano PASCOLETTI, Istituto di Matematica Informatica e Sistemistica, Università di Udine, Via Mantica 3, 33100 Udine, Italy.

Witold PEDRYCZ, Dept. of Automatic Control & Computer Sciences, Silesian Technical University, 44-100 Gliwice, Poland.

Alberto PERETTI, Centro di Informatica e Calcolo Automatico, Facoltà di Economia e Commercio, Università di Verona, Via dell'Artigliere 19, 37129 Verona, Italy.

Sixto RIOS, Universidad Complutense, Isaac Peral 1, Madrid, Spain.

Elemer E. ROSINGER, Department of Mathematics, University of Pretoria, Pretoria 0002, South Africa.

Francesco ROSSI, Istituto di Statistica, Università di Verona, Via dell'Artigliere 19, 37129 Verona, Italy.

Paolo SERAFINI, Istituto di Matematica Informatica e Sistemistica, Università di Udine, Via Mantica 3, 33100 Udine, Italy.

Andrzej M. SKULIMOWSKI, ICSET, University of Mining and Metallurgy, Al. Mickiewicza 30, 30059 Krakow, Poland.

Tetsuzo TANINO, Dept. of Mechanical Engineering II, Tohoku University, Sendai 980, Japan.

Po L. YU, School of Business, 350 Summerfield Hall, University of Kansas, Lawrence, Kansas 66045, USA.

Stanley ZIONTS, School of Management, Crosby Hall, State University of New York, Buffalo, N.Y. 14214 USA.

Printed in the United States
By Bookmasters